A Student's Guide to Numerical Methods

This concise, plain-language guide, for senior undergraduates and graduate students, aims to develop intuition, practical skills, and an understanding of the framework of numerical methods for the physical sciences and engineering. It provides accessible, self-contained explanations of mathematical principles, avoiding intimidating formal proofs. Worked examples and targeted exercises enable the student to master the realities of using numerical techniques for common needs such as the solution of ordinary and partial differential equations, fitting experimental data, and simulation using particle and Monte Carlo methods.

Topics are carefully selected and structured to build understanding, and illustrate key principles such as: accuracy, stability, order of convergence, iterative refinement, and computational effort estimation. Enrichment sections and in-depth footnotes form a springboard to more advanced material and provide additional background. Whether used for self-study, or as the basis of an accelerated introductory class, this compact textbook provides a thorough grounding in computational physics and engineering.

IAN H. HUTCHINSON is a plasma physicist and Professor in the School of Engineering at MIT where he has been solving physical problems using computers for 40 years. A fellow of the American Physical Society and the Institute of Physics, he has won several awards for excellence in teaching at MIT, and is the author of the definitive textbook on making measurements of plasmas, *Principles of Plasma Diagnostics* (Cambridge University Press, 2002).

A Student's Guide to Numerical Methods

IAN H. HUTCHINSON
Massachusetts Institute of Technology

CAMBRIDGE
UNIVERSITY PRESS

University Printing House, Cambridge CB2 8BS, United Kingdom

One Liberty Plaza, 20th Floor, New York, NY 10006, USA

477 Williamstown Road, Port Melbourne, VIC 3207, Australia

314-321, 3rd Floor, Plot 3, Splendor Forum, Jasola District Centre, New Delhi - 110025, India

79 Anson Road, #06-04/06, Singapore 079906

Cambridge University Press is part of the University of Cambridge.

It furthers the University's mission by disseminating knowledge in the pursuit of education, learning and research at the highest international levels of excellence.

www.cambridge.org
Information on this title: www.cambridge.org/9781107095670

© Cambridge University Press 2015

This publication is in copyright. Subject to statutory exception and to the provisions of relevant collective licensing agreements, no reproduction of any part may take place without the written permission of Cambridge University Press.

First published 2015

A catalogue record for this publication is available from the British Library

Library of Congress Cataloging in Publication data
Hutchinson, I. H. (Ian H.), 1951–
A student's guide to numerical methods /
I. H. Hutchinson, Massachusetts Institute of Technology.
pages cm
Includes bibliographical references and index.
ISBN 978-1-107-09567-0 (Hardback) – ISBN 978-1-107-47950-0 (Paperback)
1. Numerical analysis. 2. Measurement. 3. Number concept. 4. Matrices.
5. Mathematics–Study and teaching. I. Title.
QA39.3.H92 2015
518–dc23 2014043961

ISBN 978-1-107-09567-0 Hardback
ISBN 978-1-107-47950-0 Paperback

Cambridge University Press has no responsibility for the persistence or accuracy of URLs for external or third-party internet websites referred to in this publication, and does not guarantee that any content on such websites is, or will remain, accurate or appropriate.

S.D.G.

Contents

Preface

This book presents what every graduate-level physicist and engineer should know about solving physical problems by computer.

Hardly any research engineer or scientist, whatever their speciality, can do without at least minimal competence in computational and numerical methods. It helps the practitioner greatly to appreciate the big picture of how computational techniques are applied. A book like this that covers the breadth of the methods, with a minimum of fuss, serves the purpose of acquiring the essential knowledge. It is derived from an accelerated short course for entering graduate students in the MIT Department of Nuclear Science and Engineering. That's why some examples used to illustrate the numerical techniques are drawn from nuclear science and engineering. But no specific background nuclear knowledge is required. The mathematical and computational techniques explained are applicable throughout a whole range of engineering and physical science disciplines, because the underlying numerical methods are essentially common.

For so short a course, a great deal of background must be taken for granted, and a lot of relevant topics omitted. The brevity is not a fault though; it is an intention. And while there is an enormous range of material that *could* be added, I see the deliberate selection as a merit. This approach, I believe, enables a student to read the text sequentially, experience rapid progress, and work to *master* the content. Of course the present approach contrasts strongly both with comprehensive textbooks and with handbooks. Massive teaching *textbooks*, in addition to providing vastly more detail, cover topics such as standard matrix inversion or decomposition, and elementary quadrature. Those can mostly be taken for granted today, I believe, because of widespread use of mathematical computing systems. Large textbooks also often approach the topics by a round-about set of examples and develop the mathematics in a more elementary and drawn-out style. Doubtless that approach has merit, but it requires much more time to get to the heart of the matter. Students with good preparation appreciate a more accelerated approach than slogging through many hundreds of pages of textbook. That's especially so, since such textbooks often stop just when the techniques are getting useful for science and engineering, that is, with partial differential equations. This book gets to partial differential equations by the second quarter of its material, and continues into a discussion of particle and Monte Carlo methods that are essential for modern computational science and engineering, but rarely treated in general numerical methods textbooks. There are several excellent numerical methods *handbooks*; and I often use them. They cover so much material, though, that the reader can and should only dip into it, and use it for reference. Benefitting from such comprehensive treatments requires an understanding of the framework of the subject. The present book aims to provide that framework in as compact a form as is reasonable.

Even having deliberately chosen to keep it short, there are some places where additional explanation or details seem really valuable. To maintain the continuity and brevity of the main text, the additional material has been placed in *enrichment* sections. Enrichment material can safely be omitted (and is omitted from the lectures and the expectations of the half-term course as I teach it) but its presence gives additional background and allows an interested reader to see, still briefly, where the cited results come from. The main material (excluding the footnotes, enrichment, worked examples, or exercises) of each chapter except the last is designed to be covered in a one-and-a-half-hour lecture. But most students then need to spend time reviewing the material of that lecture, including the worked examples.

Although the course includes mathematical and computational exercises, which implement the algorithms that are discussed, it does not teach programming. A student who wants to master the material must tackle the exercises. To do so, the student must already have, or develop as they go, sufficient familiarity with a chosen language or computational system. The exercises have been tested with the use of Octave, which is an open-source alternative with practically the same syntax and capability as Matlab. Many students will find these systems appropriate, because they have built-in graphics and matrix variable types and routines, but many other language choices are possible. Elementary familiarity with matrices and the terms of linear algebra is presumed. That knowledge is reviewed in highly abbreviated form in the appendix. A principle of the development is that routine matrix algebra and functions are just used; standard direct matrix algorithms are not explained or programmed. However, iterative matrix techniques that are specific to solving differential equations are introduced on the basis of physical intuition as part of the main development, and a brief introduction to modern iterative sparse-matrix solution techniques is given in the final chapter.

Many, perhaps most, problems encountered in physical science and engineering are expressed in terms of ordinary or partial differential equations. Familiarity with vector calculus is therefore an absolute prerequisite. As much as is needed of the general theory of partial differential equations is developed within the course, but it is the merest outline. While a serious effort is made to provide self-consistent accuracy of expression and mathematics, no pretence whatever is made to mathematical rigor. There are no theorems here. The purpose is not to teach mathematical proofs; it is to equip students with the mathematical insights and tools to understand and use computational techniques. This outlook permits us to approach topics in a sequence that is intuitive, but would be ruled out by insistence on mathematical rigor.

Understanding how to solve the equations that govern common physical systems comes partly from knowing how they arise. Therefore, some are derived from first principles. The majority of students will have seen some comparable derivation in their earlier education. Therefore, the derivations here are terse. And while the text is mostly self-contained, the learning curve for students without any prior background in diffusion, fluid flow, collisions, or the kinetic theory of gases, will be rather steep. It may be advisable for them to supplement the treatment with wider reading.

Students that complete this book including the exercises will

- become familiar with computational engineering and its mathematical foundations, at an elementary level;
- deepen their understanding of the basic equations governing the physical phenomena;
- understand the methods by which problems can be solved using computation;
- develop experience, confidence, and good critical judgement in the application of numerical methods to the solution of physical problems; and
- strengthen their ability to use computation in theoretical analysis and experimental data interpretation.

The idea behind these objectives is that for students who do not specialize in computational engineering or science, this might be the *last* course they take in numerical techniques.

Though the aim is to provide what *every* physicist and engineer should know about computational problem solving, it is not to provide *everything* they should know. Graduate students who undertake computationally based research or other professional activities in science or engineering will certainly eventually need more than is covered here. For them, however, this might be a useful *first* course, because it rapidly surveys a wide range of algorithms, thus quickly providing a broad perspective on computational science technique.

Acknowledgements This book owes a great deal to students in the course Essential Numerical Methods, who by their interest, questions, corrections, and occasional puzzlement have helped me identify and explain the key conceptual challenges of learning computational physics and engineering. I am grateful to MIT for support in developing the material of this book; and always to my wife Fran, for her endless love and care, without which this work would not have happened.

1

Fitting functions to data

1.1 Exact fitting

1.1.1 Introduction

Suppose we have a set of real-number data pairs $x_i, y_i,\ i = 1, 2, \ldots, N$. These can be considered to be a set of points in the xy-plane. They can also be thought of as a set of values y of a function of x; see Fig. 1.1. A frequent challenge is to find some kind of function that represents a "best fit" to the data in some sense. If the data were fitted perfectly, then clearly the function f would have the property

$$f(x_i) = y_i, \qquad \text{for all } i = 1, \ldots, N. \tag{1.1}$$

When the number of pairs is small and they are reasonably spaced out in x, then it may be reasonable to do an exact fit that satisfies this equation.

1.1.2 Representing an exact fitting function linearly

We have an infinite choice of possible fitting functions. Those functions must have a number of different adjustable parameters that are set so as to adjust the

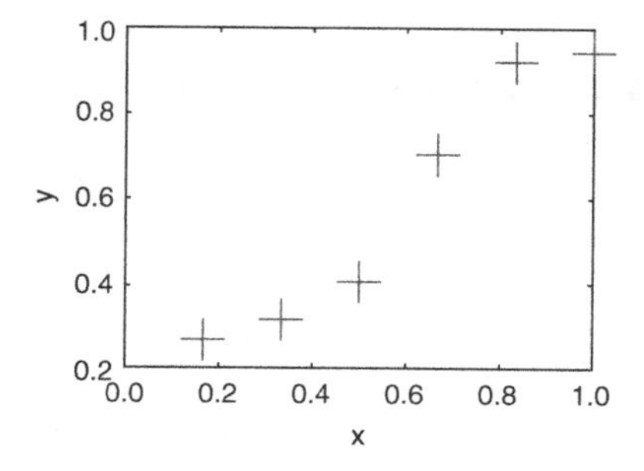

Figure 1.1 Example of data to be fitted with a curve.

function to fit the data. One example is a polynomial,

$$f(x) = c_1 + c_2 x + c_3 x^2 + \ldots + c_N x^{N-1}. \tag{1.2}$$

Here the c_i are the coefficients that must be adjusted to make the function fit the data. A polynomial whose coefficients are the adjustable parameters has a very useful property that it is linearly dependent upon the coefficients.

In order to fit eqs. (1.1) with the form of eq. (1.2) requires that N simultaneous equations be satisfied. Those equations can be written as an $N \times N$ matrix equation as follows:

$$\begin{pmatrix} 1 & x_1 & x_1^2 & \ldots & x_1^{N-1} \\ 1 & x_2 & x_2^2 & \ldots & x_2^{N-1} \\ \ldots & & & & \\ 1 & x_N & x_N^2 & \ldots & x_N^{N-1} \end{pmatrix} \begin{pmatrix} c_1 \\ c_2 \\ \ldots \\ c_N \end{pmatrix} = \begin{pmatrix} y_1 \\ y_2 \\ \ldots \\ y_N \end{pmatrix} \tag{1.3}$$

Here we notice that in order for this to be a square matrix system we need the number of coefficients to be equal to the number of data pairs N.

We also see that we could have used any set of N functions f_i as fitting functions, and written the representation:

$$f(x) = c_1 f_1(x) + c_2 f_2(x) + c_3 f_3(x) + \ldots + c_N f_N(x) \tag{1.4}$$

and then we would have obtained the matrix equation

$$\begin{pmatrix} f_1(x_1) & f_2(x_1) & f_3(x_1) & \ldots & f_N(x_1) \\ f_1(x_2) & f_2(x_2) & f_3(x_2) & \ldots & f_N(x_2) \\ \ldots & & & & \\ f_1(x_N) & f_2(x_N) & f_3(x_N) & \ldots & f_N(x_N) \end{pmatrix} \begin{pmatrix} c_1 \\ c_2 \\ \ldots \\ c_N \end{pmatrix} = \begin{pmatrix} y_1 \\ y_2 \\ \ldots \\ y_N \end{pmatrix} \tag{1.5}$$

This is the most general form of representation of a fitting function that varies linearly with the unknown coefficients. The matrix[1] we will call $\mathbf{S}$. It has elements $S_{ij} = f_j(x_i)$

1.1.3 Solving for the coefficients

When we have a matrix equation of the form $\mathbf{Sc} = \mathbf{y}$, where $\mathbf{S}$ is a square matrix, then provided that the matrix is non-singular, that is, provided its

[1] Throughout this book, matrices and vectors in abstract multidimensional space are denoted by upright bold font. Vectors in physical three-dimensional space are instead denoted by italic bold font.

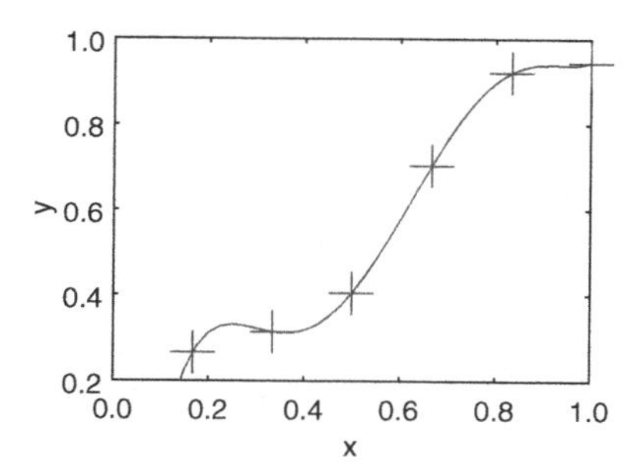

Figure 1.2 Result of the polynomial fit.

determinant is non-zero, $|\mathbf{S}| \neq 0$, it possesses an inverse $\mathbf{S}^{-1}$. Multiplying on the left by this inverse we get:

$$\mathbf{S}^{-1}\mathbf{S}\mathbf{c} = \mathbf{c} = \mathbf{S}^{-1}\mathbf{y}. \tag{1.6}$$

In other words, we can solve for $\mathbf{c}$, the unknown coefficients, by inverting the function matrix, and multiplying the values to be fitted, $\mathbf{y}$, by that inverse.

Once we have the values of $\mathbf{c}$ we can evaluate the function $f(x)$ (eq. 1.2) at any x-value we like. Fig. 1.2 shows the result of fitting a fifth order polynomial (with six terms including the 1) to the six points of our data. The line goes exactly through every point. But there's a significant problem that the line is unconvincingly curvy near its ends.[2] It's not a terribly good fit.

1.2 Approximate fitting

If we have lots of data which have scatter in them, arising from uncertainties or noise, then we almost certainly *do not* want to fit a curve so that it goes exactly through every point. For example, see Fig. 1.3. What do we do then? Well, it turns out that we can use almost exactly the same approach, except with *different* number of points (N) and terms (M) in our linear fit. In other words we use a representation

$$f(x) = c_1 f_1(x) + c_2 f_2(x) + c_3 f_3(x) + \ldots + c_M f_M(x), \tag{1.7}$$

in which usually $M < N$. We know now that we *can't* fit the data exactly. The set of equations we would have to satisfy to do so would be

[2] This problem with polynomial fitting is sometimes called Runge's phenomenon.

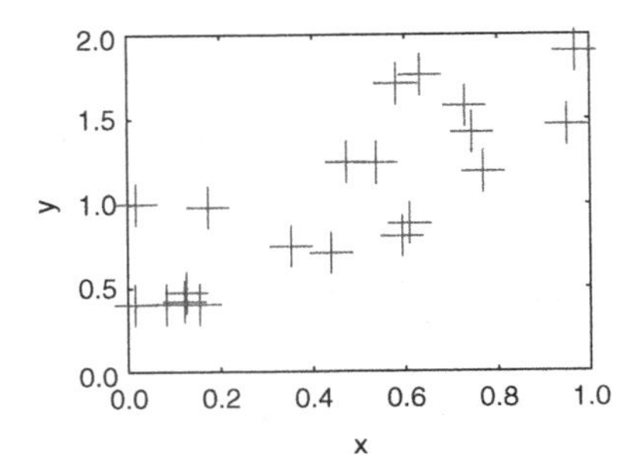

Figure 1.3 A cloud of points with uncertainties and noise, to be fitted with a function.

$$\begin{pmatrix} f_1(x_1) & f_2(x_1) & f_3(x_1) & \dots & f_M(x_1) \\ f_1(x_2) & f_2(x_2) & f_3(x_2) & \dots & f_M(x_2) \\ \dots & & & & \\ f_1(x_N) & f_2(x_N) & f_3(x_N) & \dots & f_M(x_N) \end{pmatrix} \begin{pmatrix} c_1 \\ c_2 \\ \dots \\ c_M \end{pmatrix} = \begin{pmatrix} y_1 \\ y_2 \\ \dots \\ y_N \end{pmatrix}, \quad (1.8)$$

in which the function matrix **S** is now not square but has dimensions $N \times M$. There are not enough coefficients c_j to be able to satisfy these equations exactly. They are over-specified. Moreover, a non-square matrix doesn't have an inverse.

But we are not interested in fitting these data exactly. We want to fit some sort of line through the points that best fits them.

1.2.1 Linear least squares

What do we mean by "best fit"? Especially when fitting a function of the linear form eq. (1.7), we usually mean that we want to minimize the vertical distance between the points and the line. If we had a fitted function $f(x)$, then for each data pair (x_i, y_i), the square of the vertical distance between the line and the point is $(y_i - f(x_i))^2$. So the sum, over all the points, of the square distance from the line is

$$\chi^2 = \sum_{i=1,N} (y_i - f(x_i))^2. \quad (1.9)$$

We use the square of the distances in part because they are always positive. We don't want to add positive and negative distances, because a negative distance is just as bad as a positive one and we don't want them to cancel out. We generally call χ^2 the "residual", or more simply the "chi-squared". It is an inverse measure of goodness of fit. The smaller it is the better. A linear least-squares problem is: find the coefficients of our function f that minimize the residual χ^2.

1.2.2 SVD and the Moore–Penrose pseudo-inverse

We seem to have gone off in a different direction from our original way to solve for the fitting coefficients by inverting the square matrix $\mathbf{S}$. How is that related to the finding of the least-squares solution to the over-specified set of equations (1.8)?

The answer is a piece of matrix magic! It turns out that there *is* (contrary to what one is taught in an elementary matrix course) a way to define the inverse of a non-square matrix or of a singular square matrix. It is called the (Moore–Penrose) pseudo-inverse. And once found it can be used in essentially exactly the way it was for the non-singular square matrix in the earlier treatment. That is, we solve for the coefficients using $\mathbf{c} = \mathbf{S}^{-1}\mathbf{y}$, except that $\mathbf{S}^{-1}$ is now the pseudo-inverse.

The pseudo-inverse is best understood from a consideration of what is called the singular value decomposition (SVD) of a matrix. This is the embodiment of a theorem in matrix mathematics that states that any $N \times M$ matrix can always be expressed as the product of three other matrices with very special properties. For our $N \times M$ matrix $\mathbf{S}$ this expression is:

$$\mathbf{S} = \mathbf{U}\mathbf{D}\mathbf{V}^T, \tag{1.10}$$

where ^{T} denotes transpose, and

- $\mathbf{U}$ is an $N \times N$ orthonormal matrix
- $\mathbf{V}$ is an $M \times M$ orthonormal matrix
- $\mathbf{D}$ is an $N \times M$ diagonal matrix.

Orthonormal[3] means that the dot product of any column (regarded as a vector) with any other column is zero, and the dot product of a column with itself is unity. The inverse of an orthonormal matrix is its transpose. So

$$\underbrace{\mathbf{U}^T}_{N\times N}\underbrace{\mathbf{U}}_{N\times N} = \underbrace{\mathbf{I}}_{N\times N} \quad \text{and} \quad \underbrace{\mathbf{V}^T}_{M\times M}\underbrace{\mathbf{V}}_{M\times M} = \underbrace{\mathbf{I}}_{M\times M}, \tag{1.11}$$

A diagonal matrix has non-zero elements only on the diagonal. But if it is non-square, as it is if $M < N$, then it is padded with extra rows of zeros (or extra columns if $N < M$):

[3] Sometimes called simply "orthogonal," the real version of "unitary."

$$\mathbf{D} = \begin{pmatrix} d_1 & 0 & 0 & \dots & 0 \\ 0 & d_2 & 0 & & \\ 0 & & \ddots & & \\ \vdots & & & \ddots & 0 \\ & & 0 & 0 & d_M \\ 0 & 0 & 0 & 0 & 0 \end{pmatrix}. \tag{1.12}$$

A sense of what the SVD is can be gained from by thinking[4] in terms of the eigenanalysis of the matrix $\mathbf{S}^T\mathbf{S}$. Its eigenvalues are d_i^2.

The pseudo-inverse can be considered to be

$$\mathbf{S}^{-1} = \mathbf{V}\mathbf{D}^{-1}\mathbf{U}^T. \tag{1.13}$$

Here, $\mathbf{D}^{-1}$ is an $M \times N$ diagonal matrix whose entries are the inverse of those of $\mathbf{D}$, i.e. $1/d_j$:

$$\mathbf{D}^{-1} = \begin{pmatrix} 1/d_1 & 0 & 0 & \dots & 0 & 0 \\ 0 & 1/d_2 & 0 & & & 0 \\ 0 & & \ddots & & & 0 \\ \vdots & & & \ddots & 0 & 0 \\ 0 & \dots & 0 & 0 & 1/d_M & 0 \end{pmatrix}. \tag{1.14}$$

It's clear that eq. (1.13) is in some sense an inverse of $\mathbf{S}$ because formally

$$\mathbf{S}^{-1}\mathbf{S} = (\mathbf{V}\mathbf{D}^{-1}\mathbf{U}^T)(\mathbf{U}\mathbf{D}\mathbf{V}^T) = \mathbf{V}\mathbf{D}^{-1}\mathbf{D}\mathbf{V}^T = \mathbf{V}\mathbf{V}^T = \mathbf{I}. \tag{1.15}$$

[4] **Enrichment:** A highly abbreviated outline of the SVD is as follows. The $M \times M$ matrix $\mathbf{S}^T\mathbf{S}$ is symmetric. Therefore, it has real eigenvalues d_i^2, which because of its form are non-negative. Its eigenvectors $\boldsymbol{v}_i$, satisfying $\mathbf{S}^T\mathbf{S}\boldsymbol{v}_i = d_i^2\boldsymbol{v}_i$, can be arranged into an orthonormal set in order of decreasing magnitude of d_i^2. The M eigenvectors can be considered the columns of an orthonormal matrix $\mathbf{V}$, which diagonalizes $\mathbf{S}^T\mathbf{S}$ so that $\mathbf{V}^T\mathbf{S}^T\mathbf{S}\mathbf{V} = \mathbf{D}^2$ is an $M \times M$ non-negative, diagonal matrix with diagonal values d_i^2. Since $(\mathbf{S}\mathbf{V})^T\mathbf{S}\mathbf{V}$ is diagonal, the columns of the $N \times M$ matrix $\mathbf{S}\mathbf{V}$ are orthogonal. Its columns corresponding to $d_i = 0$ are zero and are not useful to us. The useful columns corresponding to non-zero d_i ($i = 1, \dots, L$, say, $L \leq M$) can be normalized by dividing by d_i. Then, by appending $N - L$ normalized column N-vectors that are orthogonal to all the previous ones, we can construct a complete $N \times N$ orthonormal matrix $\mathbf{U} = [\mathbf{S}\mathbf{V}\mathbf{D}_L^{-1}, \mathbf{U}_{N-L}]$. Here $\mathbf{D}_L^{-1}$ denotes the $M \times L$ diagonal matrix with elements $1/d_i$ and $\mathbf{U}_{N-L}$ denotes the appended extra columns. Now consider $\mathbf{U}\mathbf{D}\mathbf{V}^T$. The appended $\mathbf{U}_{N-L}$ make zero contribution to this product because the lower rows of $\mathbf{D}$ which they multiply are always zero. The rest of the product is $\mathbf{S}\mathbf{V}\mathbf{D}_L^{-1}\mathbf{D}_L\mathbf{V}^T = \mathbf{S}$. Therefore we have constructed the singular value decomposition $\mathbf{S} = \mathbf{U}\mathbf{D}\mathbf{V}^T$.

If $M \leq N$ and none of the d_j is zero, then all the operations in this matrix multiplication reduction are valid, because

$$\underbrace{\mathbf{D}^{-1}}_{M\times N}\underbrace{\mathbf{D}}_{N\times M} = \underbrace{\mathbf{I}}_{M\times M}. \tag{1.16}$$

But see the enrichment section[5] for a detailed discussion of other cases.

The most important thing for our present purposes is that if $M \leq N$ then we can find a solution of the over-specified (rectangular matrix) fitting problem $\mathbf{Sc} = \mathbf{y}$ as $\mathbf{c} = \mathbf{S}^{-1}\mathbf{y}$, using the pseudo-inverse. The set of coefficients $\mathbf{c}$ we get corresponds to more than one possible set of y_i-values, but that does not matter.

Also, it can be shown[6], that the specific solution that is obtained by this matrix product is in fact the *least-squares solution* for $\mathbf{c}$; i.e. the solution that minimizes the residual χ^2. And if there is any freedom in the choice of $\mathbf{c}$, such that the residual is at its minimum for a range of different $\mathbf{c}$, then the solution which minimizes $|\mathbf{c}|^2$ is the one found.

The beauty of this fact is that one can implement a simple code, which calls a function `pinv` to find the pseudo-inverse, and it will work just fine if the matrix $\mathbf{S}$ is singular or even rectangular.

As a matter of computational efficiency, it should be noted that in Octave the backslash operator is equivalent to multiplying by the pseudo-inverse (i.e. `pinv(S)*y = S\y`), but is calculated far more efficiently.[7] So backslash is preferable in computationally costly code, because it is roughly five times

[5] **Enrichment:** If $M > N$, the combination $\mathbf{DD}^{-1}$, which arises from forming $\mathbf{SS}^{-1}$, is not an $M \times M$ identity matrix. Instead it has ones only, at most, for the first N of the diagonal positions, and zeros thereafter. It is an $N \times N$ identity matrix with extra zero rows and columns padding it out to $M \times M$. So the pseudo-inverse is a funny kind of inverse, which works only one way.

If $\mathbf{S}$ is square and *non-singular*, then the pseudo-inverse is exactly the same as the (normal) inverse.

If $\mathbf{S}$ were a *singular* square matrix, for example (and possibly in other situations), then at least one of the original d_j would be zero. We usually consider the singular values (d_j) to be arranged in descending order of size; so that the zero values come at the end. $\mathbf{D}^{-1}$ would then have an element $1/d_j$ that is *infinite*, and the formal manipulations would be unjustified. What the pseudo-inverse does in these tricky cases is put the value of the inverse $1/d_j$ equal to zero instead of infinity. In that case, once again, an incomplete identity matrix is produced, with extra diagonal zeros at the end. And it actually doesn't completely "work" as an inverse in either direction.

For those who know some linear algebra, what's happening is that the pseudo-inverse projects vectors in the range of the original matrix back into the complement of its nullspace.

[6] See for example, first edition W. H. Press, B. P. Flannery, S. A. Teukolsky, and W. T. Vettering (1989), *Numerical Recipes*, Cambridge University Press, Cambridge, (henceforth cited simply as *Numerical Recipes*), Section 2.9.

[7] By *QR* decomposition.

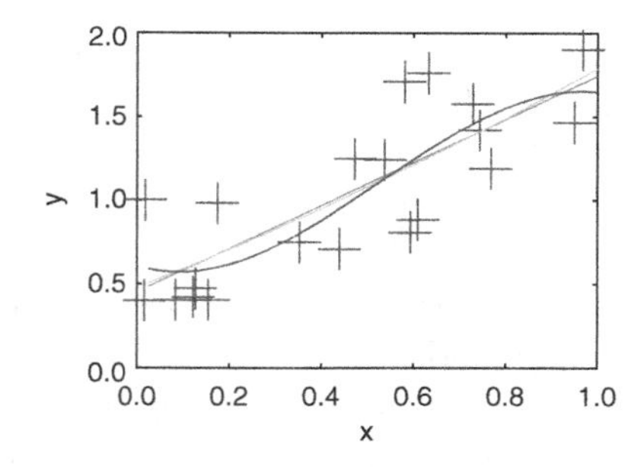

Figure 1.4 The cloud of points fitted with linear, quadratic, and cubic polynomials.

faster. You probably won't notice the difference for matrix dimensions smaller than a few hundred.

1.2.3 Smoothing and regularization

As we illustrate in Fig. 1.4, by choosing the number of degrees of freedom of the fitting function the smoothness of the fit can be adjusted to the data. However, the choice of basis functions then constrains one in a way that has been pre-specified. It might not in fact be the best way to smooth the data to fit them by (say) a straight line or a parabola.

A better way to smooth is by "regularization" in which we add some measure of roughness to the residual we are seeking to minimize. The roughness (which is the inverse of the smoothness) is a measure of how wiggly the fit line is. It can in principle be pretty much anything that can be written in the form of a matrix times the fit coefficients. I'll give an example in a moment. Let's assume the roughness measure is homogeneous, in the sense that we are trying to make it as near zero as possible. Such a target would be $\mathbf{Rc} = 0$, where $\mathbf{R}$ is a matrix of dimension $N_R \times M$, where N_R is the number of distinct roughness constraints. Presumably we can't satisfy this equation perfectly because a fully smooth function would have no variation, and be unable to fit the data. But we want to minimize the square of the roughness $(\mathbf{Rc})^T\mathbf{Rc}$. We can try to fulfil the requirement to fit the data, and to minimize the roughness, in a least-squares sense by constructing an expanded compound matrix system combining the original equations and the regularization; thus: [8]

[8] This notation means the first N rows of the compound matrix consist of $\mathbf{S}$, and the next N_R rows are $\lambda\mathbf{R}$.

$$\begin{pmatrix} \mathbf{S} \\ \lambda\mathbf{R} \end{pmatrix} \mathbf{c} = \begin{pmatrix} \mathbf{y} \\ \mathbf{0} \end{pmatrix}. \tag{1.17}$$

If we solve this system in a least-squares sense by using the pseudo-inverse of the compound matrix $\begin{pmatrix} \mathbf{S} \\ \lambda\mathbf{R} \end{pmatrix}$, then we will have found the coefficients that "best" make the roughness zero as well as fitting the data: in the sense that the total residual

$$\chi^2 = \sum_{i=1,N} (y_i - f(x_i))^2 + \lambda^2 \sum_{k=1,N_R} \left(\sum_j R_{kj} c_j \right)^2 \tag{1.18}$$

is minimized. The value of λ controls the weight of the smoothing. If it is large, then we prefer smoother solutions. If it is small or zero, we do negligible smoothing.

As a specific one-dimensional example, we might decide that the roughness we want to minimize is represented by the second derivative of the function: d^2f/dx^2. Making this quantity on average small has the effect of minimizing the wiggles in the function, so it is an appropriate roughness measure. We could therefore choose $\mathbf{R}$ such that it represented that derivative at a set of chosen points x_k, $k = 1, N_R$ (not the same as the data points x_i) in which case:

$$R_{kj} = \left. \frac{d^2 f_j}{dx^2} \right|_{x_k}. \tag{1.19}$$

The x_k might, for example, be equally spaced over the x-interval of interest, in which case[9] the squared roughness measure could be considered a discrete approximation to the integral, over the interval, of the quantity $(d^2f/dx^2)^2$.

1.3 Tomographic image reconstruction

Consider the problem of x-ray tomography. We make many measurements of the integrated density of matter along chords in a plane section through some object whose interior we wish to reconstruct. These are generally done by measuring the attenuation of x-rays along each chord, but the mathematical technique is independent of the physics. We seek a representation of the density

[9] This regularization is equivalent to what is sometimes called a "smoothing spline." In the limit of large smoothing parameter λ, the function f is a straight line (zero second derivative everywhere). In the limit of small λ, it is a cubic spline interpolation through all the values (x_i, y_i).

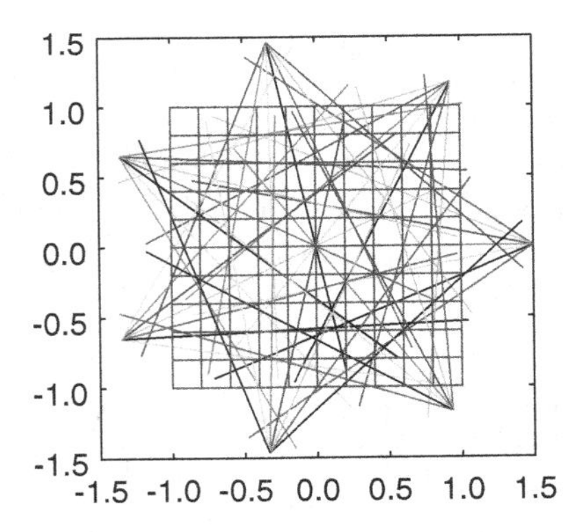

Figure 1.5 Illustrative layout of tomographic reconstruction of density in a plane using multiple fans of chordal observations.

of the object in the form

$$\rho(x,y) = \sum_{j=1,M} c_j \rho_j(x,y), \tag{1.20}$$

where $\rho_j(x,y)$ are basis functions over the plane. They might actually be as simple as pixels over mesh x_k and y_l, such that $\rho_j(x,y) \to \rho_{kl}(x,y) = 1$ when $x_k < x < x_{k+1}$ and $y_l < y < y_{l+1}$, and zero otherwise. However, the form of basis function that won A. M. Cormack the Nobel prize for medicine in his implementation of "computerized tomography" (the CT scan) was much more cleverly chosen to build the smoothing into the basis functions. Be careful thinking about multidimensional fitting. For constructing fitting matrices, the list of basis functions should be considered to be logically arranged from 1 to M in a single index j so that the coefficients are a single column vector. But the physical arrangement of the basis functions might more naturally be expressed using two indices k, l referring to the different spatial dimensions. If so then they must be mapped in some consistent manner to the vector column.

Each chord along which measurements are made passes through the basis functions (e.g. the pixels), and for a particular set of coefficients c_j we therefore get a chordal measurement value

$$v_i = \int_{l_i} \rho d\ell = \int_{l_i} \sum_{j=1,M} c_j \rho_j(x,y) d\ell = \sum_{j=1,M} \int_{l_i} \rho_j(x,y) d\ell \; c_j = \mathbf{Sc}, \tag{1.21}$$

where the $N \times M$ matrix $\mathbf{S}$ is formed from the integrals along each of the N lines of sight l_i, so that $S_{ij} = \int_{l_i} \rho_j(x,y) d\ell$. It represents the contribution of basis function j to measurement i. Our fitting problem is thus rendered into the

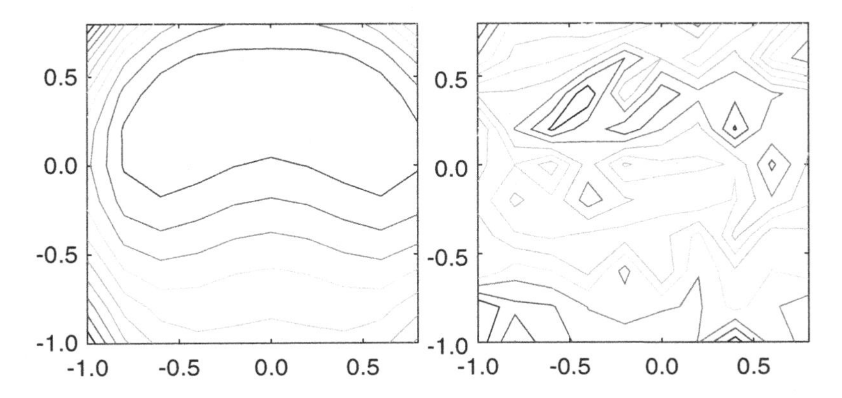

Figure 1.6 Contour plots of the initial test ρ-function (left) used to calculate the chordal integrals, and its reconstruction based upon inversion of the chordal data (right). The number of pixels (100) exceeds the number of views (49), and the number of singular values used in the pseudo-inverse is restricted to 30. Still they do not agree well, because various artifacts appear. Reducing the number of singular values does not help.

standard form:

$$\mathbf{S}\mathbf{c} = \mathbf{v}, \tag{1.22}$$

in which a rather large number M of basis functions might be involved. We can solve this by pseudo-inverse: $\mathbf{c} = \mathbf{S}^{-1}\mathbf{v}$, and, if the system is overdetermined such that the effective number of different chords is larger than the number of basis functions, it will probably work.

The problem is, however, usually *underdetermined*, in the sense that we don't really have enough independent chordal measurements to determine the density in each pixel (for example). This is true even if we apparently have more measurements than pixels, because generally there is a finite noise or uncertainty level in the chordal measurements that becomes amplified by the inversion process. This is illustrated by a simple test as shown in Fig. 1.6.

We then almost certainly want to smooth the representation otherwise all sorts of meaningless artifacts will appear in our reconstruction that have no physical existence. If we try to do this by forming a pseudo-inverse in which a smaller number of singular values are retained, and the others put to zero, there is no guarantee that this will get rid of the roughness. Fig. 1.6 gives an example.

If instead we smooth the reconstruction by regularization, using as our measure of roughness the discrete (two-dimensional) Laplacian ($\nabla^2\rho$) evaluated at each pixel, we get a far better result, as shown in Fig. 1.7. It turns out that this

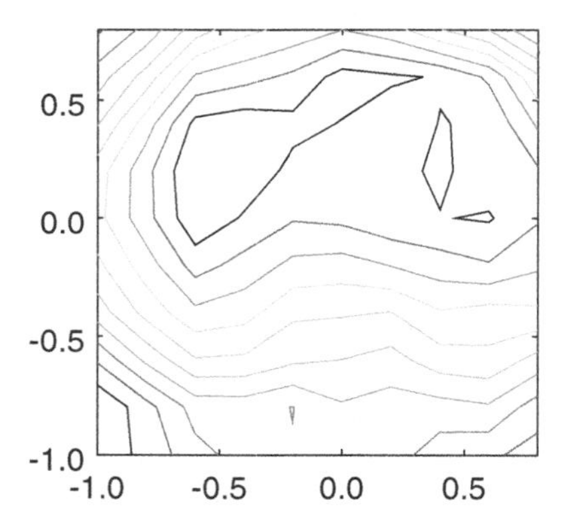

Figure 1.7 Reconstruction using a regularization smoothing based upon $\nabla^2\rho$. The contours are much nearer to reality.

good result is rather insensitive to the value of λ^2 over two or three orders of magnitude.

1.4 Efficiency and non-linearity

Using the inverse or pseudo-inverse to solve for the coefficients of a fitting function is intuitive and straightforward. However, in many cases it is *not* the most computationally efficient approach. For moderate size problems, modern computers have more than enough power to overcome the inefficiencies, but in a situation with multiple dimensions, such as tomography, it is easy for the matrix that needs to be inverted to become enormous, because that matrix's side length is the *total number* of pixels or elements in the fit, which may be, for example, the product of the side lengths `nx`×`ny`. The giant matrix that has to be inverted may be very "sparse," meaning that all but a very few of its elements are zero. It can then become overwhelming in terms of storage and cpu to use the direct inversion methods we have discussed here. We'll see other approaches later.

Some fitting problems are *non-linear*. For example, suppose one had a photon spectrum of a particular spectral line to which one wished to fit a Gaussian function of particular center, width, and height. That's a problem that cannot be expressed as a linear sum of functions. In that case fitting becomes more elaborate,[10] and less reliable. There are some potted fitting programs out there, but it's usually better if you can avoid them.

[10] This topic, and many others in data fitting, is addressed for example in S. Brandt (2014), *Data Analysis: Statistical and Computational Methods for Scientists and Engineers*, fourth edition, Springer, New York.

Worked example. Fitting sinusoidal functions

Suppose we wish to fit a set of data x_i, y_i spread over the range of independent variables $a \leq x \leq b$. And suppose we know the function is zero at the boundaries of the range, at $x = a$ and $x = b$. It makes sense to incorporate our knowledge of the boundary values into the choice of functions to fit, and choose those functions f_n to be zero at $x = a$ and $x = b$. There are numerous well-known sets of functions that have the property of being zero at two separated points. The points where standard functions are zero are of course not some arbitrary a and b. But we can scale the independent variable x so that a and b are mapped to the appropriate points for any choice of function set.

Suppose the functions that we decide to use for fitting are sinusoids:[11] $f_n = \sin(n\theta)$ all of which are zero at $\theta = 0$ and $\theta = \pi$. We can make this set fit our x range by using the scaling

$$\theta = \pi(x - a)/(b - a), \tag{1.23}$$

so that θ ranges from 0 to π as x ranges from a to b. Now we want to find the best fit to our data in the form

$$f(x) = c_1 \sin(\theta) + c_2 \sin(2\theta) + c_3 \sin(3\theta) + \cdots + c_M \sin(M\theta). \tag{1.24}$$

We therefore want the least-squares solution for the c_i of

$$\mathbf{Sc} = \begin{pmatrix} \sin(1\theta_1) & \sin(2\theta_1) & \dots & \sin(M\theta_1) \\ \sin(1\theta_2) & \sin(2\theta_2) & \dots & \sin(M\theta_2) \\ \dots & & & \\ \sin(1\theta_N) & \sin(2\theta_N) & \dots & \sin(M\theta_N) \end{pmatrix} \begin{pmatrix} c_1 \\ c_2 \\ \dots \\ c_M \end{pmatrix} = \begin{pmatrix} y_1 \\ y_2 \\ \dots \\ y_N \end{pmatrix} = \mathbf{y} \tag{1.25}$$

We find this solution by the following procedure.

1. If necessary, construct column vectors $\mathbf{x}$ and $\mathbf{y}$ from the data.
2. Calculate the scaled vector $\boldsymbol{\theta}$ from $\mathbf{x}$.
3. Construct the matrix $\mathbf{S}$ whose ijth entry is $\sin(j\theta_i)$
4. Least-squares-solve $\mathbf{Sc} = \mathbf{y}$ (e.g. by pseudo-inverse) to find $\mathbf{c}$.
5. Evaluate the fit at any x by substituting the expression for θ, eq. (1.23), into eq. (1.24).

This process may be programmed in a mathematical system like Matlab or Octave, which has built-in matrix multiplication, very concisely[12] as follows (entries following % are comments).

[11] Of course Fourier analysis is usually approached differently, as briefly discussed in the final chapter. We are here just using sine as an example function, in part because it is familiar.

[12] There are lots of other correct but more verbose ways of doing it.

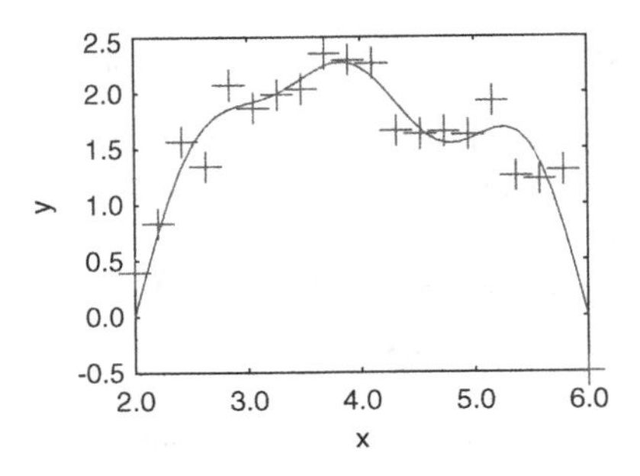

Figure 1.8 The result of the fit of sinusoids up to $M = 5$ to a noisy dataset of size $N = 20$. The points are the input data. The curve is constructed by using the `yfit` expression on an `xfit` array of some convenient length spanning the x-range, and then simply plotting `yfit` versus `xfit`.

```
% Suppose x and y exist as column vectors of length N. (Nx1 matrices)
j=[1:M];                 % Create a 1xM matrix containing numbers 1 to M.
theta=pi*(x-a)/(b-a);    % Scale x to obtain the column vector theta.
S=sin(theta*j);          % Construct the matrix S using an outer product.
Sinv=pinv(S);            % Pseudo invert it.
c=Sinv*y;                % Matrix multiply y to find the coefficients c.
```

The fit can then be evaluated for any x value (or array) `xfit`, in the form effectively of a scalar product of $\sin(\theta \mathbf{j})$ with $\mathbf{c}$. The code is likewise astonishingly brief, and will need careful thought (especially noting what the dimensions of the matrices are) to understand what is actually happening:

```
yfit=sin(pi*(xfit-a)/(b-a)*j)*c; % Evaluate the yfit at any xfit.
```

An example is shown in Fig. 1.8.

Exercise 1. Data fitting

1. Given a set of N values y_i of a function $y(x)$ at the positions x_i, write a short code to fit a polynomial having order one less than N (so there are N coefficients of the polynomial) to the data.

Obtain a set of (N =) 6 numbers from http://silas.psfc.mit.edu/22.15/15numbers.html (or if that is not accessible use $y_i = [0.892, 1.44, 1.31, 1.66, 1.10, 1.19]$). Take the values y_i to be at the positions $x_i = [0.0, 0.2, 0.4, 0.6, 0.8, 1.0]$. Run your code on these data and find the coefficients c_j.

Plot together (on the same plot) the resulting fitted polynomial representing $y(x)$ (with sufficient resolution to give a smooth curve) and the original data points, over the domain $0 \leq x \leq 1$.

Submit the following as your solution:

1. Your code in a computer format that is capable of being executed.
2. The numeric values of your coefficients c_j, $j = 1, N$.
3. Your plot.
4. Brief commentary ($<$ 300 words) on what problems you faced and how you solved them.

2. Save your code from part 1. Make a copy of it with a new name and change the new code as needed to fit (in the linear least-squares sense) a polynomial of order possibly lower than $N - 1$ to a set of data x_i, y_i (for which the points are in no particular order).

Obtain a pair of data sets of length ($N =$) 20 numbers x_i, y_i from the same URL by changing the entry in the "Number of Numbers" box. (Or, if that is inaccessible, generate your own data set from random numbers added to a line.) Run your code on these data to produce the fitting coefficients c_j when the number of coefficients of the polynomial is ($M =$) (a) 1, (b) 2, (c) 3. That is: constant, linear, quadratic.

Plot the fitted curves and the original data points on the same plot(s) for all three cases.

Submit the following as your solution:

1. Your code in a computer format that is capable of being executed.
2. Your coefficients c_j, $j = 1, M$, for three cases (a), (b), (c).
3. Your plot(s).
4. Very brief remarks on the extent to which the coefficients are the same for the three cases.
5. Can your code from this part also solve the problem of part 1?

2

Ordinary differential equations

2.1 Reduction to first order

An ordinary differential equation involves just one *independent* variable, x, and a *dependent* variable y. Obviously it involves *derivatives* of the dependent variable like $\frac{dy}{dx}$. The highest order differential, i.e. the term $\frac{d^N y}{dx^N}$ with the largest value of N appearing in the equation, defines the *order* N of the equation. So the most general ODE of order N can be written such that the Nth order derivative is equal to a function of all the lower order derivatives and the independent variable x:

$$\frac{d^N y}{dx^N} = f\left(\frac{d^{N-1}y}{dx^{N-1}}, \frac{d^{N-2}y}{dx^{N-2}}, \ldots, \frac{dy}{dx}, y, x\right). \tag{2.1}$$

Such an ordinary differential equation of order N in a single dependent variable y can always be reduced to a set of simultaneous coupled *first*-order equations involving N dependent variables. The simplest way to do this is to use a natural notation to denote by $y^{(i)}$ the ith derivative:

$$\frac{d^i y}{dx^i} = y^{(i)} \qquad i = 1, 2, \ldots, N-1. \tag{2.2}$$

When combined with the original equation, the total system can be written as a first-order vector differential equation whose components are

$$\frac{d}{dx}y^{(i)} = f_i(y^{(N-1)}, y^{(N-2)}, \ldots, y^{(1)}, y^{(0)}, x), \quad i = 0, 1, \ldots, N-1, \tag{2.3}$$

(where for notational consistency $y^{(0)} = y$). Explicitly in vector form:

$$\frac{d}{dx}\begin{pmatrix} y^{(0)} \\ y^{(1)} \\ \vdots \\ y^{(N-1)} \end{pmatrix} = \begin{pmatrix} f_0 \\ f_1 \\ \vdots \\ f_{N-1} \end{pmatrix} = \begin{pmatrix} y^{(1)} \\ y^{(2)} \\ \vdots \\ f(y^{(N-1)}, y^{(N-2)}, \ldots, y^{(1)}, y^{(0)}, x) \end{pmatrix}. \tag{2.4}$$

Recognizing that the combined simultaneous vector system of dimension N with first-order derivatives is equivalent to a single scalar equation of order N, we often say that the order of the coupled vector system is still N. (Sorry if that seems confusing. In practice you get the hang of it.)

This is formal mathematics and applies to all equations, but precisely such a set of coupled first-order equations will often also arise directly in the formulation of the practical problem we are trying to solve. Suppose we are trying to track the position of a fluid element in a three-dimensional steady flow. If we know the fluid velocity $\boldsymbol{v}$ as a function of position $\boldsymbol{v}(\boldsymbol{x})$, then the equation of the track of a fluid element, i.e. the path followed by the element as it moves in time t, is

$$\frac{d}{dt}\boldsymbol{x} = \boldsymbol{v}. \tag{2.5}$$

This is the equation we must solve to find a fluid streamline. It is of just the same form we derived by order reduction. Such a history of position as a function of time is called generically an orbit. Here the independent variable is t, and the dependent variable is $\boldsymbol{x}$. The vector $\boldsymbol{v}$ plays the role of the functions f_i.

Orbits may not be just in space, they may be in higher-dimensional phase-space. For example (see Fig. 2.1) to find the orbit of a charged particle (e.g. proton) of mass m_p and charge e moving at velocity $\boldsymbol{v}$ in a uniform magnetic field $\boldsymbol{B}$, we observe that it is subject to a force $e\boldsymbol{v} \times \boldsymbol{B}$. In the absence of any other forces, it has an equation of motion

$$\frac{d}{dt}\boldsymbol{v} = \frac{e}{m_p}\boldsymbol{v} \times \boldsymbol{B} = \boldsymbol{f}(\boldsymbol{v}), \tag{2.6}$$

in which the acceleration depends upon the velocity. This is a first-order vector differential equation, in three dimensions, where t plays the role of the

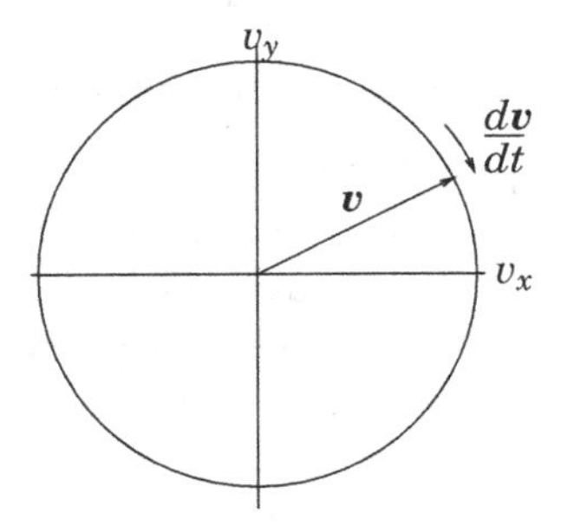

Figure 2.1 Orbit of the velocity of a particle moving in a uniform magnetic field is a circle in *velocity-space* perpendicular to the field ($B = B\hat{z}$ here).

independent variable, and the dependent variable is the vector velocity $\boldsymbol{v}$. The vector acceleration $\boldsymbol{f}$, which is the vector derivative function, depends upon all the components of $\boldsymbol{v}$.

If, for our proton orbit, $\boldsymbol{B}$ is not uniform, but varies with position, then we need to know both position $\boldsymbol{x}$ and velocity $\boldsymbol{v}$ at all times along the orbit to solve it. The system then involves six-dimensional vectors consisting of the components of $\boldsymbol{x}$ and $\boldsymbol{v}$:

$$\frac{d}{dt}\begin{pmatrix} x_1 \\ x_2 \\ x_3 \\ v_1 \\ v_2 \\ v_3 \end{pmatrix} = \begin{pmatrix} v_1 \\ v_2 \\ v_3 \\ (v_2 B_3 - v_3 B_2)e/m_p \\ (v_3 B_1 - v_1 B_3)e/m_p \\ (v_1 B_2 - v_2 B_1)e/m_p \end{pmatrix}. \tag{2.7}$$

Very often, to find *analytic* solutions it feels more natural to eliminate some of the dependent variables with the result that the order of the ODE is raised. So, for example for a uniform magnetic field in the 3-direction, the dynamics perpendicular to it separate out into

$$\frac{d}{dt}v_1 = \Omega v_2,\ \frac{d}{dt}v_2 = -\Omega v_1 \quad \Rightarrow \quad \frac{d^2 v_1}{dt^2} = -\Omega^2 v_1,\ \frac{d^2 v_2}{dt^2} = -\Omega^2 v_2 \tag{2.8}$$

(writing $\Omega = eB/m_p$). The second-order *uncoupled* equations are familiar to us as simple harmonic oscillator equations, having solutions like $\cos \Omega t$ and $\sin \Omega t$. So they are easier to solve *analytically*. But the original first-order equations, even though they are *coupled*, are far easier to solve *numerically*. So we don't generally do the elimination in computational solutions.

2.2 Numerical integration of initial-value problems

2.2.1 Explicit integration

Now we consider how in practice to solve a first-order ordinary differential equation in which all the boundary conditions are imposed at the same position in the independent variable. Such boundary conditions constitute what is called an "initial-value problem." We start integrating forward in the independent variable (e.g. time or space) from a place where the initial values are specified. To simplify the discussion we will consider a single (scalar) dependent variable y, but note that the generalization to a vector of dependent variables is usually immediate, so the treatment is fine for higher-order equations that have been reduced to vectorial first-order form.

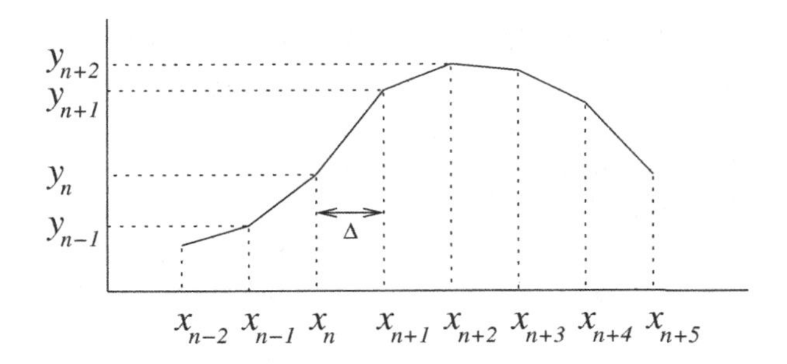

Figure 2.2 Illustrating finite difference representation of a continuous function.

In general, numerical solution of differential equations requires us to represent the solution, which is usually continuous, in a discrete manner where the values are given at a series of points rather than continuously. See Fig. 2.2. The natural way to discretize the derivative is to write

$$\frac{dy}{dx} \approx \frac{y_{n+1} - y_n}{x_{n+1} - x_n} = f(y, x), \tag{2.9}$$

where the index n denotes the value at the nth discrete step, and therefore

$$y_{n+1} = y_n + f(y, x)(x_{n+1} - x_n). \tag{2.10}$$

This equation tells us how y changes from one step to the next. Starting from an initial position we can step discretely as far as we like, choosing the size of the independent variable step $(x_{n+1} - x_n) = \Delta$ appropriately.

A question that arises, though, is what to use for x and y inside the derivative function $f(y, x)$. The x value can be chosen more or less at will[1] but before we've actually made the step, we don't know where we are going to end up in y, so we can't easily decide where in y to evaluate f. The easiest answer, but not generally the best, is to recognize that at any point in stepping from n to $n + 1$ along the orbit, we already have the value y_n. So we could just use $f(y_n, x_n)$. This choice is said to be "explicit," and is sometimes called the Euler method. The reason why this method is not the best is because it tends to have poor *accuracy* and poor *stability*.

[1] If f is independent of y, then we should use $f(x_{n+1/2})$ where $x_{n+1/2} = (x_{n+1} + x_n)/2$, and we then have a formula accurate to second order for performing simple integration $\int_0^{x_n} f(x)dx = y_n - y_0 = \sum_{k=0}^{n-1} f(x_{k+1/2})(x_{k+1} - x_k)$. There are fancier, higher-order, ways to do integration, but this is by far the most straightforward.

2.2.2 Accuracy and Runge–Kutta schemes

To illustrate the problem of accuracy, consider the derivative function f expanded as a Taylor series about the x_n, y_n position, writing $x - x_n = \delta x$, $y - y_n = \delta y$. The derivative function f is a function of both x and y. However, the solution for the orbit can be written $y = y(x)$. Therefore, the function evaluated on the orbit, $f(y(x), x)$, is a function only of x, and we can write its (total) derivative as $\frac{df}{dx}$. The Taylor expansion of this function is simply[2]

$$f(y(x), x) = f(y_n, x_n) + \frac{df_n}{dx}\delta x + \frac{d^2 f_n}{dx^2}\frac{\delta x^2}{2!} + O(\delta x^3). \tag{2.11}$$

We use the notation df_n/dx (etc.) to indicate values evaluated at position n. If we substitute this Taylor expansion for f into the differential equation we are trying to solve, $dy/dx = d\delta y/d\delta x = f$, and integrate term by term, we get the *exact* solution of the differential equation:

$$\delta y = f_n \delta x + \frac{df_n}{dx}\frac{\delta x^2}{2!} + \frac{d^2 f_n}{dx^2}\frac{\delta x^3}{3!} + O(\delta x^4). \tag{2.12}$$

We subtract from it whatever the finite difference *approximate* equation is. In the case of eq. (2.10) it is $\delta y^{(1)} = f_n \delta x$, and we find that the error in y_{n+1} is

$$\delta y - \delta y^{(1)} = \epsilon = \frac{df_n}{dx}\frac{\delta x^2}{2!} + \frac{d^2 f_n}{dx^2}\frac{\delta x^3}{3!} + O(\delta x^4). \tag{2.13}$$

This tells us that the explicit Euler difference scheme is accurate only to *first order* in the size of the step δx (when the first derivative of f is non-zero) because an error of order δx^2 is present. That means if we make the step a factor of two smaller, the cumulative error, when integrating over a set total distance, gets smaller by (approximately) a factor of two. (Because each step's error is four times smaller, but there are twice as many steps.) That's not very good. We would have to take very small steps, δx, to get good accuracy.

We can do better. Our error arose because we approximated the derivative function by using its value only at x_n. But once we've moved to the next position, and know (with some inaccuracy) the value y_{n+1} and hence f_{n+1} there, we can evaluate *better* the f we should have used. This process is illustrated in Fig. 2.3. In fact, by substitution from eq. (2.11) it is easy to see that if we use, instead, the advancing equation

$$\delta y = \frac{1}{2}(f_n + f_{n+1}^{(1)})\delta x, \tag{2.14}$$

[2] The notation $O(\epsilon^n)$ denotes additional terms whose magnitude is proportional to a (usually small) parameter ϵ raised to the power n or higher.

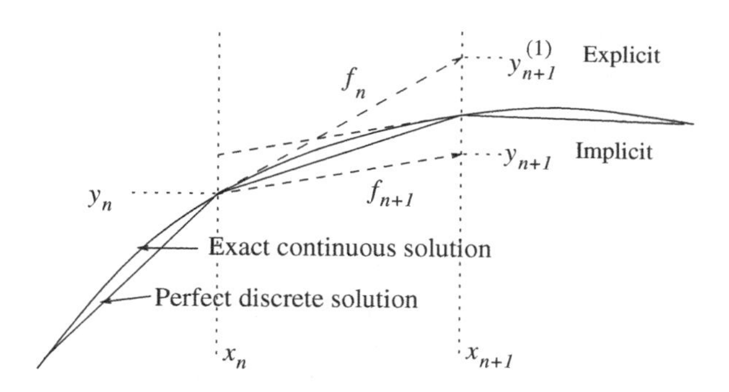

Figure 2.3 Optional steps using derivative function evaluated at n or $n+1$.

where $f_{n+1}^{(1)} = f(y_{n+1}^{(1)}, x_{n+1})$ is the value of f obtained at the end of our first (explicit Euler) advance $y_{n+1}^{(1)} = y_n + f_n \delta x$, then we would obtain for our *approximate* advancing scheme

$$\delta y = f_n \delta x + \frac{df_n}{dx} \frac{\delta x^2}{2!} + O(\delta x^3), \tag{2.15}$$

which now agrees with the *first two* terms of the full exact expansion (2.12), and whose error, in comparison with that expression, is now of *third* order, rather than second. This second-order accurate scheme gives cumulative errors proportional to δx^2 and so converges much more quickly as we shorten the step length.

The reason we obtained a more accurate step was that we used a more accurate value for the average (over the step) of the derivative function. It is straightforward to improve the average even more, so as to obtain even higher-order accuracy. But to do that requires us to obtain estimates of the derivative function part way through the step as well as at the ends. That's because we need to estimate the first and second derivatives of f.

A Runge–Kutta method consists of taking a series of steps, each one of which uses the estimate of the derivative function obtained from the previous one, and then taking some weighted average of their derivatives. Specifically, the *fourth-order (accurate) Runge–Kutta* scheme, which is by far the most popular and is illustrated in Fig. 2.4, uses:

$$\begin{aligned}
f^{(0)} &= f(y_n, x_n) \\
f^{(1)} &= f\left(y_n + f^{(0)}\frac{\Delta}{2}, x_n + \frac{\Delta}{2}\right) \\
f^{(2)} &= f\left(y_n + f^{(1)}\frac{\Delta}{2}, x_n + \frac{\Delta}{2}\right) \\
f^{(3)} &= f(y_n + f^{(2)}\Delta, x_n + \Delta),
\end{aligned} \tag{2.16}$$

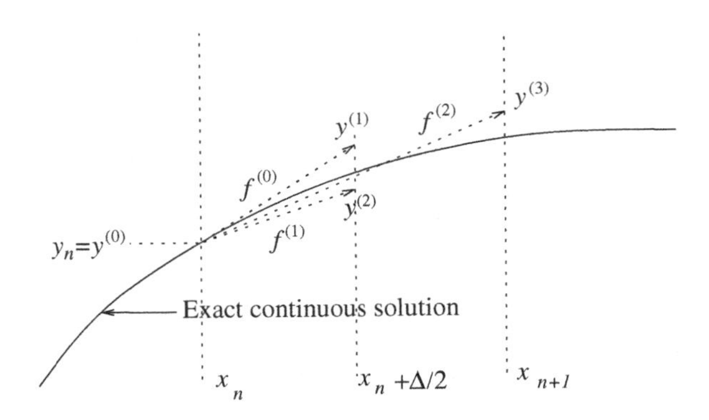

Figure 2.4 Runge–Kutta fourth-order scheme with four partial steps, evaluates the derivative function $f^{(k)}$ with $k = 0, 1, 2, 3$, at four places (x_k, y_k), each determined by extrapolation along the previous derivative.

in which two steps are to the half-way point $\frac{\Delta}{2}$. Then the following combination,

$$y_{n+1} = y_n + \left(\frac{f^{(0)}}{6} + \frac{f^{(1)}}{3} + \frac{f^{(2)}}{3} + \frac{f^{(3)}}{6} \right) \Delta + O(\Delta^5), \tag{2.17}$$

gives an approximation accurate to fourth order.[3]

The Runge–Kutta method costs more computation per step, because it requires four evaluations of the function $f(y, x)$, rather than just one. But that

[3] **Enrichment:** The proof is rather hard work. Here's an outline. Use notation $F(\delta x) = f(y(x_n + \delta x), \delta x)$ to refer to values of the derivative function along the exact orbit. Suppose we compose $\Delta y = (\frac{1}{6}F(0) + \frac{1}{3}F(\frac{\Delta}{2}) + \frac{1}{3}F(\frac{\Delta}{2}) + \frac{1}{6}F(\Delta))\Delta$. This is the form corresponding to eqs. (2.16) and (2.17), but using the exact f on the orbit, rather than the intermediate $f^{(n)}$ estimates. By substituting from eq. (2.11) it is easy but tedious to show that this Δy is equal to $y(x_n + \Delta) - y_n + O(\Delta^5)$; that is, it is an accurate representation of the y-step to fourth order. Actually it is instructive to realize that the symmetry of the δx positions, 0, $\frac{\Delta}{2}$, and Δ ensures that all even-order errors in this Δy are zero while the first-order error is zero because the coefficients sum to unity. The factor of two difference between the center and the end coefficients is the choice that makes the third-order error zero. In itself this has not proved the scheme to be fourth-order accurate, because there are non-zero differences approximately proportional to $\frac{\partial f}{\partial y}$ between the f-values and the F-values: $f^{(1)} - F(\frac{\Delta}{2}) = \frac{\partial f}{\partial y}(y^{(1)} - y(\frac{\Delta}{2})), f^{(2)} - F(\frac{\Delta}{2}) = \frac{\partial f}{\partial y}(y^{(2)} - y(\frac{\Delta}{2}))$, and $f^{(3)} - F(\Delta) = \frac{\partial f}{\partial y}(y^{(3)} - y(\Delta))$. (Because of the order of the y-differences one can quickly see that $\frac{\partial^2}{\partial y^2}$ terms don't need to be kept.) The successive differences such as $(y^{(1)} - y(\frac{\Delta}{2}))$ can be expressed in terms of the Taylor expansion eq. (2.12), and then the f-differences are gathered in the combination of eq. (2.17): $\frac{1}{6}, \frac{1}{3}, \frac{1}{3}, \frac{1}{6}$. One can then evaluate the $\frac{\partial f}{\partial y}$ terms and demonstrate that they cancel to fourth order.

is often more than compensated by the ability to take larger steps than with the Euler method for the same accuracy.

2.2.3 Stability

The second, and possibly more important, weakness of explicit integration is in respect of stability. Consider a linear differential equation

$$\frac{dy}{dx} = -ky, \tag{2.18}$$

where k is a positive constant. This of course has the solution $y = y_0 \exp(-kx)$. But suppose we integrate it numerically using the explicit scheme

$$y_{n+1} = y_n + f(y_n, x_n)(x_{n+1} - x_n) = y_n(1 - k\Delta). \tag{2.19}$$

This finite difference equation has the solution

$$y_n = y_0(1 - k\Delta)^n, \tag{2.20}$$

as may be verified by the simple observation that $y_{n+1}/y_n = (1 - k\Delta)$. (This ratio is called the amplification factor.) If $k\Delta$ is a small number, then no difficulties will arise, and our scheme will produce approximately correct results. However, a choice $k\Delta > 2$ compromises not only accuracy but also *stability*. The resulting solution has alternating sign; it oscillates; but also its magnitude *increases* with n and will tend to infinity at large x. It has become unstable, as illustrated in Fig. 2.5.

In general, an explicit discrete advancing scheme requires the step in the independent variable to be less than some value (in this case $2/k$) in order to achieve stability.

An *implicit* advancing scheme, by contrast, is one in which the value of the derivative used to advance the variable is taken at the *end* of the step rather than at the *beginning*. For our example equation, this would be a numerical scheme of the form

$$y_{n+1} = y_n + f(y_{n+1}, x_{n+1})(x_{n+1} - x_n) = y_n - ky_{n+1}\Delta. \tag{2.21}$$

It is easy to rearrange this equation into

$$y_{n+1}(1 + k\Delta) = y_n. \tag{2.22}$$

This has solution

$$y_n = y_0(1 + k\Delta)^{-n}. \tag{2.23}$$

For positive $k\Delta$ (the case of interest) this finite difference equation *never* becomes unstable, no matter how big $k\Delta$ is, because the solution consists of

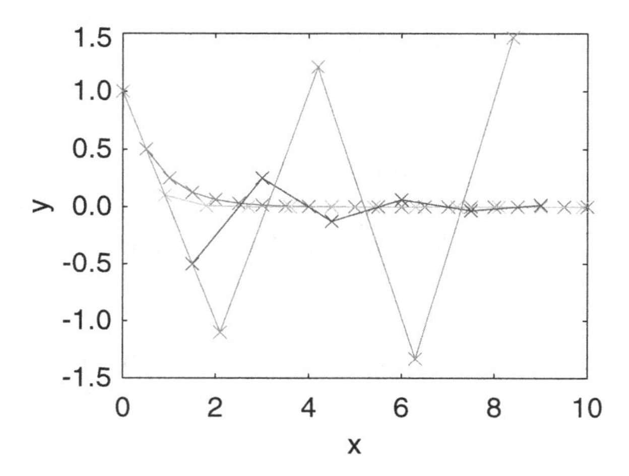

Figure 2.5 Explicit numerical integration of eq. (2.18), using eq. (2.19) leads to an oscillatory instability if the step length is too long. Four step lengths are shown $\Delta/k = 0.5, 0.9, 1.5, 2.1$.

successive powers of an amplification factor $1/(1 + k\Delta)$ whose magnitude is always less than 1. This is a characteristic of implicit schemes. They are generally stable even for large steps.

2.3 Multidimensional stiff equations: implicit schemes

The question of stability for an order-one system (the scalar problem) is generally not very interesting; because instability of the explicit scheme occurs only when the step size is longer than the characteristic spatial scale of the problem $1/k$. If you've chosen your step size so large, you are already failing to get an accurate solution of the equation. However, multi-dimensional (i.e. higher order) sets of (vector) equations may have multiple solutions that have very different scale-lengths in the independent variable. A classic example is an order-two homogeneous linear system with constant coefficients

$$\frac{d}{dx}\mathbf{y} = \mathbf{A}\mathbf{y}, \quad \text{where for example} \quad \mathbf{A} = \begin{pmatrix} 0 & -1 \\ 100 & -101 \end{pmatrix}. \tag{2.24}$$

For any such linear system the solution can be constructed by consideration of the *eigenvalues* of the matrix $\mathbf{A}$: those numbers for which there exists a solution to $\mathbf{Ay} = \lambda\mathbf{y}$. If these are λ_j and the corresponding eigenvectors are $\mathbf{y}_j$, then $\mathbf{y} = \mathbf{y}_j \exp(\lambda_j x)$ are solutions to the equation. The complete solution can be constructed as a sum of these different characteristic solutions, weighted by coefficients to satisfy the initial conditions. The point of our particular example matrix is that its eignvalues are -100 and -1. Consequently, in order to integrate numerically a solution that has a significant quantity of the second, slowly changing, solution $\lambda_2 = -1$, it is necessary nevertheless to ensure the stability of the first, rapidly changing, solution, $\lambda_1 = -100$. Otherwise, if the first solution is unstable, no matter how little of that solution we start with, it will eventually grow exponentially large and erroneously dominate our result. If an explicit advancing scheme is used, then stability requires $|\lambda_1|\Delta < 2$ as well as $|\lambda_2|\Delta < 2$, and the λ_1 condition is by far the most restrictive. There are then at least $\sim |\lambda_1/\lambda_2|$ steps during the decay of the (λ_2) solution of interest. Because this ratio is large, an explicit scheme is computationally expensive, requiring many steps. In general, the *stiffness* of a differential equation system can be measured by the ratio of the largest to the smallest eigenvalue magnitude. If this ratio is large, the system is stiff, and that means it is hard to integrate explicitly.

Using an implicit scheme avoids the necessity for taking very small steps. It does so at the cost of solving the matrix problem $(\mathbf{I} - \Delta.\mathbf{A})\mathbf{y}_{n+1} = \mathbf{y}_n$. This requires the inversion of a matrix in order to evaluate

$$\mathbf{y}_{n+1} = (\mathbf{I} - \Delta.\mathbf{A})^{-1}\mathbf{y}_n. \tag{2.25}$$

For a linear problem like the one we are considering, the single inversion, done once and for all, is a relatively small cost compared with the gain obtained by being able to take decent length steps.

All of this probably seems rather elaborate for a linear, constant coefficient, system, since we are actually able to solve it analytically when we know the eigenvalues and eigenvectors. However, it becomes much more significant when we realize that the stability of a *non-linear* system, or one in which the coefficients vary with x or y, for which numerical integration may be essential, is generally very well described by expressing it approximately as a linear system in the neighborhood of the region under consideration, and then evaluating the stability of the linearized system. The matrix that arises from linearization when the (vectorial) derivative function is $\mathbf{f}(\mathbf{y}, x)$ has the elements $A_{ij} = \partial f_i/\partial y_j$. An implicit solution then requires $(\mathbf{I} - \Delta.\mathbf{A})$ to be inverted for every step, because it is changing with position (if the derivative function is non-linear).

In short, implicit schemes lead to greater stability, which is very important with stiff systems, but they require matrix inversion.

2.4 Leap-Frog schemes

Codes such as particle-in-cell simulations of plasmas, atomistic simulation, or any codes that do a large amount of orbit integration, generally want to use as cheap a scheme as possible to maintain their speed. The step size is often dictated by questions other than the accuracy of the orbit integration. In such situations, Runge–Kutta or implicit schemes are rarely used. Instead, an accurate orbit integration can be obtained by recognizing that Newton's second law of motion (acceleration proportional to force) is a second-order vector equation that can most conveniently be split into two first-order vector equations involving position, velocity, and acceleration. The velocity we want for the equation governing the evolution of position, $d\boldsymbol{x}/dt = \boldsymbol{v}$, is the average velocity during the motion between two positions. An unbiassed estimate of that velocity is to take it to be the velocity at the *center* of the position step. So if the times at which we evaluate the position $\boldsymbol{x}_n$ of the particle are denoted t_n, then we want the velocity at time $(t_{n+1} + t_n)/2$. We might call this time $t_{n+1/2}$ and the velocity $\boldsymbol{v}_{n+1/2}$. Also, the acceleration we want for the equation for the evolution of velocity, $d\boldsymbol{v}/dt = \boldsymbol{a}$, is the average acceleration between two velocities. If the velocities are represented at $t_{n-1/2}$ and $t_{n+1/2}$, then the time at which we want the acceleration is t_n, and we might call that acceleration $\boldsymbol{a}_n$.

We can therefore construct an appropriate centered integration scheme as follows starting from position n.

(1) Move the particles using their velocities $\boldsymbol{v}_{n+1/2}$ to find the new positions $\boldsymbol{x}_{n+1} = \boldsymbol{x}_n + \boldsymbol{v}_{n+1/2}(t_{n+1} - t_n)$; this is called the drift step.
(2) Accelerate the velocities using the accelerations $\boldsymbol{a}_{n+1}$ evaluated at the new positions $\boldsymbol{x}_{n+1}$ to obtain the new velocities $\boldsymbol{v}_{n+3/2} = \boldsymbol{v}_{n+1/2} + \boldsymbol{a}_{n+1}(t_{n+3/2} - t_{n+1/2})$; this is called the kick step.
(3) Repeat (1) and (2) for the next time step $n + 1$, and so on.

Each of the drift and kick steps can be simple explicit advances. But because the velocities are suitably time-shifted relative to the positions and accelerations, the result is *second-order accurate*. Such a scheme is called a Leap-Frog scheme, in reference to the children's game where each of two players alternately jumps over the back of the other in moving to the new position. Velocity and position are jumping over one another in time; they never land at the same time (or place). See Fig. 2.6.

One trap for the unwary in a Leap-Frog scheme is the specification of initial values. If we want to calculate an orbit which has specified initial position $\boldsymbol{x}_0$ and velocity $\boldsymbol{v}_0$ at time $t = t_0$, then it is not sufficient simply to put the velocity initially equal to the specified $\boldsymbol{v}_0$. That is because the first velocity,

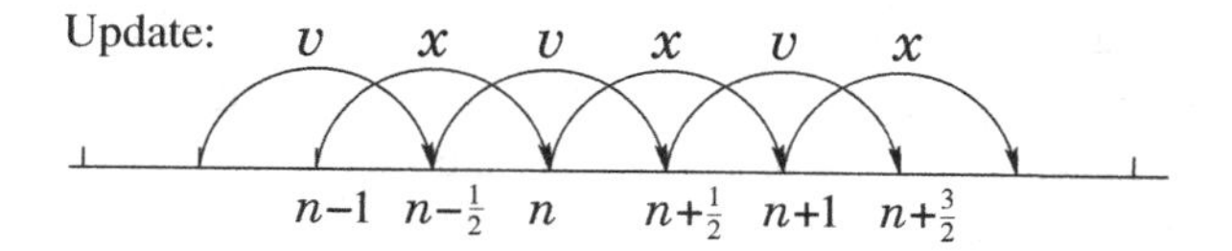

Figure 2.6 Staggered values for x and v are updated leap-frog style.

which governs the position step from t_0 to t_1 is *not* $\boldsymbol{v}_0$, but $\boldsymbol{v}_{1/2}$. To start the integration off correctly, therefore, we must take a *half-step* in velocity by putting $\boldsymbol{v}_{1/2} = \boldsymbol{v}_0 + \boldsymbol{a}_0(t_1 - t_0)/2$, before beginning the standard integration.

Leap-Frog schemes generally possess important conservation properties such as conservation of momentum, that can be essential for realistic simulation with a large number of particles.

Worked example. Stability of finite differences

Find the stability limits when solving by discrete finite differences of length Δ, using explicit or implicit schemes, the initial-value equation

$$\frac{d^2y}{dx^2} + 2\alpha\frac{dy}{dx} + k^2 y = g(x), \tag{2.26}$$

where α is a positive constant $< |k|$.

First, reduce the equation to standard first-order form by writing

$$\begin{aligned} \frac{d}{dx}\begin{pmatrix} y^{(0)} \\ y^{(1)} \end{pmatrix} &= \begin{pmatrix} y^{(1)} \\ g(x) - 2\alpha y^{(1)} - k^2 y^{(0)} \end{pmatrix} \\ &= \begin{pmatrix} 0 & 1 \\ -k^2 & -2\alpha \end{pmatrix}\begin{pmatrix} y^{(0)} \\ y^{(1)} \end{pmatrix} + \begin{pmatrix} 0 \\ g(x) \end{pmatrix}. \end{aligned} \tag{2.27}$$

Now, notice that, in either form, this equation has a sort of driving term $g(x)$ independent of y. It can be removed from the stability analysis by considering a new variable z, which is the difference between the finite difference numerical solution that we are trying to analyse, $y_n(x)$, and the exact solution of the differential equation, $y(x)$. Thus $z_n = y_n - y(x_n)$. The quantity z satisfies a discretized form of the *homogeneous* version of eq. (2.26), with the term $g(x)$ removed. Consequently analysing that homogeneous equation tells us whether the difference z between the numerical and the exact solutions is stable or not. That is what we really want to know. For stability, we pay no attention to $g(x)$. The resulting homogeneous equation is of exactly the same form as eq. (2.24): $\frac{d}{dx}\mathbf{z} = \mathbf{A}\mathbf{z}$. Each independent solution for $\mathbf{z}$ is thus an eigenvector of the matrix

$\mathbf{A}$ such that $\frac{d}{dx}\mathbf{z} = \lambda\mathbf{z}$. The eigenvalue λ decides its stability. An explicit (Euler) scheme of step length Δ will result in a step amplification factor, such that $\mathbf{z}_{n+1} = F\mathbf{z}_n$ with $F = 1 + \lambda\Delta$, and will be unstable if $|F| > 1$.

The eigenvalues of $\mathbf{A}$ satisfy

$$0 = \begin{vmatrix} -\lambda & 1 \\ -k^2 & -\lambda - 2\alpha \end{vmatrix} = \lambda^2 + 2\alpha\lambda + k^2, \tag{2.28}$$

with solutions

$$\lambda = -\alpha \pm \sqrt{\alpha^2 - k^2}. \tag{2.29}$$

These solutions are complex if $k^2 > \alpha^2$, in which case,

$$|F|^2 = (1 - \alpha\Delta)^2 + (k^2 - \alpha^2)\Delta^2 = 1 - 2\alpha\Delta + k^2\Delta^2. \tag{2.30}$$

$|F|$ is greater than unity unless $\Delta < 2\alpha/k^2$, which is the stability criterion of the Euler scheme. If $\alpha = 0$, so that the homogeneous equation is an undamped harmonic oscillator, the explicit scheme is unstable for any Δ. A fully implicit scheme, by contrast, has $F = \mathbf{z}_{n+1}/\mathbf{z}_n = 1/(1 - \lambda\Delta)$, for which $|F|^2 = 1/(1 + 2\alpha\Delta + k^2)$, always less than one. The implicit scheme is always stable.

Exercise 2. Integrating ordinary differential equations

1. Reduce the following higher-order ordinary differential equations to first-order vector differential equations, which you should write out in vector format:

(a) $\frac{d^2y}{dt^2} = -1$,

(b) $Ay + B\frac{dy}{dx} + C\frac{d^2y}{dx^2} + D\frac{d^3y}{dx^2} = E$, and

(c) $\frac{d^2y}{dx^2} = 2\left(\frac{dy}{dx}\right)^2 - y^3$.

2. Accuracy order of ODE schemes. For notational convenience, we start at $x = y = 0$ and consider a small step in x and y of the ODE $dy/dx = f(y, x)$. The Taylor expansion of the derivative function along the orbit is

$$f(y(x), x) = f_0 + \frac{df_0}{dx}x + \frac{d^2f_0}{dx^2}\frac{x^2}{2!} + \dots. \tag{2.31}$$

(a) Integrate $\frac{dy}{dx} = f(y(x), x)$ term by term to find the solution for y to third order in x.

(b) Suppose $y_1 = f_0 x$. Find $y_1 - y(x)$ to second order in x.

(c) Now consider $y_2 = f(y_1, x)x$, show that it is equal to $f(y, x)x$ plus a term that is third order in x.

(d) Hence find $y_2 - y$ to second order in x.
(e) Finally, show that $y_3 = \frac{1}{2}(y_1 + y_2)$ is equal to y accurate to *second order* in x.

[Comment. The third-order term in part (c) involves the partial derivative $\partial f/\partial y$ rather than the derivative along the orbit. Proving rigorously that the fourth-order Runge–Kutta scheme really is fourth order is rather difficult because it requires keeping track of such partial derivatives.]

3. **Programming exercise.** Write a program to integrate numerically from $t = 0$ to $t = 4/\omega$ the ODE

$$\frac{dy}{dt} = -\omega y,$$

with ω a positive constant, starting from $y(0) = 1$, proceeding as follows.

(a) Use the explicit Euler scheme

$$y_{n+1} = y_n - \Delta t \omega y_n.$$

(b) Use the implicit scheme

$$y_{n+1} = y_n - \Delta t \omega y_{n+1}.$$

In each case, find numerically the *fractional* error at $t = 4/\omega$ for the following choices of timestep:

(i) $\omega\Delta t = 0.1$, (ii) $\omega\Delta t = 0.01$, and (iii) $\omega\Delta t = 1$.

(c) Find experimentally the timestep value at which the explicit scheme becomes unstable. Verify that the implicit scheme never becomes unstable.

3

Two-point boundary conditions

Very often, the boundary conditions that determine the solution of an ordinary differential equation are applied not just at a single value of the independent variable, x, but at two points, x_1 and x_2. This type of problem is inherently different from the "initial-value problems" discussed previously. Initial-value problems are single-point boundary conditions. There must be more than one *condition* if the system is higher order than one, but in an initial-value problem, all conditions are applied at the same place (or time). In two-point problems we have boundary conditions at more than one *place* (more than one value of the independent variable) and we are interested in solving for the dependent variable(s) in the interval $x_1 \leq x \leq x_2$ of the independent variable.

3.1 Examples of two-point problems

Many examples of two-point problems arise from steady-flux conservation in the presence of sources.

In electrostatics the electric potential ϕ is related to the charge density ρ through one of the Maxwell equations: a form of Poisson's equation

$$\nabla.\boldsymbol{E} = -\nabla^2\phi = \rho/\epsilon_0, \tag{3.1}$$

where $\boldsymbol{E}$ is the electric field and ϵ_0 is the permittivity of free space. In a slab geometry where ρ varies in a single direction (coordinate) x, but not in y or z, an ordinary differential equation arises

$$\frac{d^2\phi}{dx^2} = -\frac{\rho(x)}{\epsilon_0}. \tag{3.2}$$

If we suppose (see Fig. 3.1) that the potential is held equal to zero at two planes, x_1 and x_2, by placing grounded conductors there, then its variation between

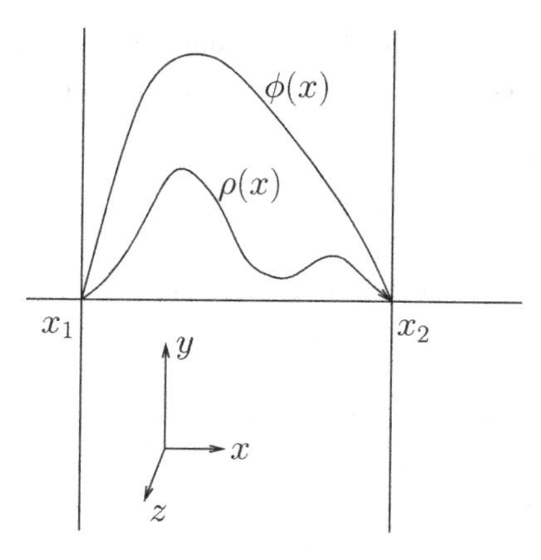

Figure 3.1 Electrostatic configuration independent of y and z with conducting boundaries at x_1 and x_2 where $\phi = 0$. This is a second-order two-point boundary problem.

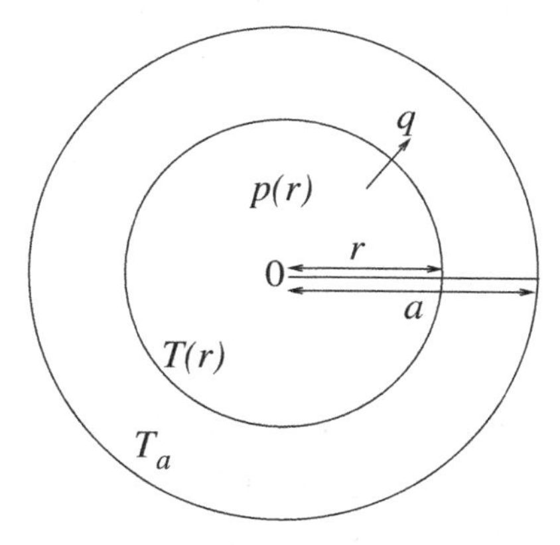

Figure 3.2 Heat balance equation in cylindrical geometry leads to a two-point problem with conditions consisting of fixed temperature at the edge, $r = a$ and zero gradient at the center $r = 0$.

them depends upon the distribution of the charge density $\rho(x)$. Solving for $\phi(x)$ is a two-point problem. In effect, this is a conservation equation for electric flux, $\boldsymbol{E}$. Its divergence is equal to the source density, which is the charge density.

A second-order two-point problem also arises from steady heat conduction. See Fig. 3.2. Suppose a cylindrical reactor fuel rod experiences volumetric heating from the nuclear reactions inside it with a power density $p(r)$ (watts per cubic meter) that varies with cylindrical radius r.

Its boundary, at $r = a$ say, is held at a constant temperature T_a. If the thermal conductivity of the rod is $\kappa(r)$, then the radial heat flux density (watts per square meter) is

$$q = -\kappa \frac{dT}{dr}. \tag{3.3}$$

In steady state, the total heat flux across the surface at radius r (per unit rod length) must equal the total heating within it:

$$2\pi r q = -2\pi r \kappa(r) \frac{dT}{dr} = \int_0^r p(r') 2\pi r' dr'. \tag{3.4}$$

Differentiating this equation we obtain:

$$\frac{d}{dr}\left(r\kappa \frac{dT}{dr} \right) = -rp(r). \tag{3.5}$$

This second-order differential equation requires two boundary conditions. One is $T(a) = T_a$, but the other is less immediately obvious. It is that the solution must be non-singular at $r = 0$, which requires that the *derivative* of T be zero there:

$$\left.\frac{dT}{dr}\right|_{r=0} = 0. \tag{3.6}$$

3.2 Shooting

3.2.1 Solving two-point problems by initial-value iteration

One approach to computing the solution to two-point problems is to use the same technique used to solve initial-value problems. We treat x_1 as if it were the starting point of an initial-value problem. We choose enough boundary conditions there to specify the entire solution. For a second-order equation such as (3.2) or (3.5), we would need to choose two conditions: $y(x_1) = y_1$, and $dy/dx|_{x_1} = s$, say, where y_1 and s are the chosen values. Only one of these is actually the boundary condition to be applied at the initial point, x_1. We'll suppose it is y_1. The other, s, is an arbitrary guess at the start of our solution procedure.

Given these initial conditions, we can solve for y over the entire range $x_1 \leq x \leq x_2$. When we have done so for this case, we can find the value y at x_2 (or its derivative if the original boundary conditions there required it). Generally, this first solution will *not* satisfy the actual two-point boundary condition at x_2, which we'll take as $y = y_2$. That's because our guess of s was

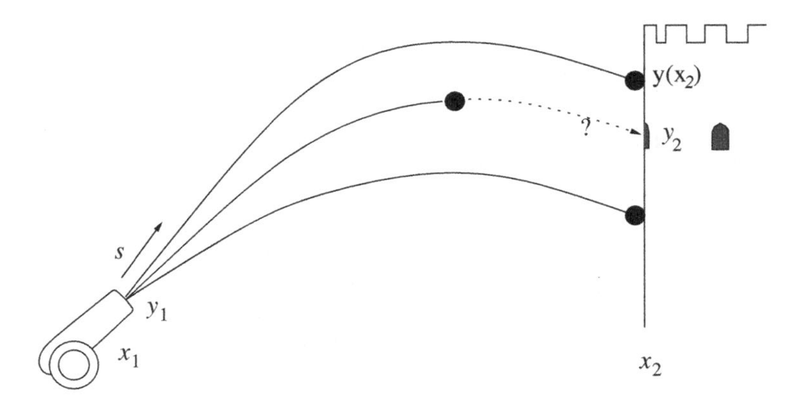

Figure 3.3 Multiple successive shots from a cannon can take advantage of observations of where the earlier ones hit, in order to iterate the aiming elevation s until they strike the target.

not correct. It's as if we are aiming at the point (x_2, y_2) with a cannon located at (x_1, y_1) (see Fig. 3.3). We elevate the cannon so that the cannonball's initial angle is $dy/dx|_{x_1} = s$, which is our initial guess at the best aim. We shoot. The cannonball flies over (within our metaphor, the initial value solution is found) but is not at the correct height when it reaches x_2 because our first guess at the aim was imperfect. What do we do? We see the height at which the cannonball hits, above or below the target. We adjust our aim accordingly with a new elevation s_2, $dy/dx|_{x_1} = s_2$, and shoot again. Then we iteratively refine our aim taking as many shots as necessary, and improving the aim each time, till we hit the target. This is the "shooting" method of solving a two-point problem. The cannonball's trajectory stands for the initial-value integration with assumed initial condition.

One question that is left open in this description is exactly *how* we refine our aim. That is, how do we change the guess of the initial slope s so as to get a solution that is nearer to the correct value of $y(x_2)$? One of the easiest and most robust ways to do this is by bisection.

3.2.2 Bisection

Suppose we have a continuous function $f(s)$ over some interval $[s_l, s_u]$ (i.e. $s_l \leq s \leq s_u$), and we wish to find a solution to $f(s) = 0$ within that range. If $f(s_l)$ and $f(s_u)$ have *opposite signs*, then we know that there is a solution (a "root" of the equation) somewhere between s_l and s_u. For definiteness in our description, we'll take $f(s_l) \leq 0$ and $f(s_u) \geq 0$. To get a better estimate

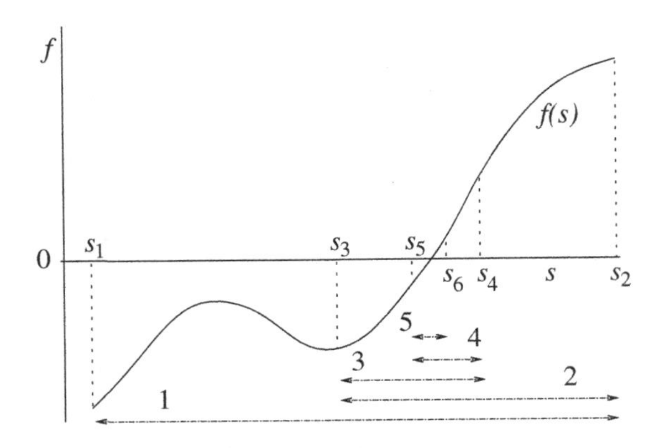

Figure 3.4 Bisection successively divides in two an interval in which there is a root, always retaining the subinterval in which a root lies.

of where $f = 0$ is, we can bisect the interval and examine the value of f at the point $s = (s_l + s_u)/2$. If $f(s) < 0$, then we know that a solution must be in the half interval $[s, s_u]$, whereas if $f(s) > 0$, then a solution must be in the other interval $[s_l, s]$. We choose whichever half-interval the solution lies in, and update one or other end of our interval to be the new s-value. In other words, we set either $s_l = s$ or $s_u = s$, respectively. The new interval $[s_l, s_u]$ is half the length of the original, so it gives us a better estimate of where the solution value is.

Now we just iterate the above procedure, as illustrated in Fig. 3.4. At each step we get an interval of half the length of the previous step, in which we know a solution lies. Eventually the interval becomes small enough that its extent can be ignored; we then know the solution accurately enough, and can stop the iteration.

The wonderful thing about bisection is that it is highly efficient, because it is guaranteed to converge in "logarithmic time." If we start with an interval of length L, then at the kth interation the interval length is $L/2^k$. So if the tolerance with which we need the s-value solution is δ (generally a small length), the number of iterations we must take before convergence is $N = \log_2(L/\delta)$. For example if $L/\delta = 10^6$, then $N = 20$. This is a quite modest number of iterations even for a very high degree of refinement.

There are iterative methods of root finding that converge faster than bisection for well behaved functions. One is "Newton's method," which may succinctly be stated as $s_{k+1} = s_k - f(s_k)/ df/ds|_{s_k}$. It converges in a few steps when the starting guess is not too far from the solution. Unlike bisection, it does not require two starting points on opposite sides of the root. However,

Newton's method (1) requires derivatives of the function, which makes it more complicated to code, and (2) is less robust, because it takes big steps near $df/ds = 0$, and may even step in the wrong direction and not converge at all in some cases. Bisection is guaranteed to converge after a modest number of steps. Robustness is in practice usually more important than speed.[1]

In the context of our shooting solution of a two-point problem, the function f is the error in the boundary value at the second point $y(x_2) - y_2$ of the inital-value solution $y(x)$ that takes initial value s for its derivative at x_1. The bisection generally adjusts the initial value s until $|y(x_2) - y_2|$ is less than some tolerance (rather than requiring some tolerance on s).

3.3 Direct solution

The shooting method, while sometimes useful for situations where adaptive step length is a major benefit, is rather a back-handed way of solving two-point problems. It is very often better to solve the problem by constructing a finite difference system to represent the differential equation *including* its boundary conditions, and then solve that system directly.

3.3.1 Second-order finite differences

First let's consider how one ought to represent a second-order derivative as finite differences. Suppose we have a uniformly spaced grid (or mesh) of values of the independent variable x_n such that $x_{n+1} - x_n = \Delta x$. The natural definition of the first derivative is

$$\left.\frac{dy}{dx}\right|_{n+1/2} = \frac{\Delta y}{\Delta x} = \frac{y_{n+1} - y_n}{x_{n+1} - x_n}; \tag{3.7}$$

and this should be regarded as an estimate of the value at the midpoint $x_{n+1/2} = (x_n + x_{n+1})/2$, which we denote via a half-integral index $n+1/2$. The second derivative is the derivative of the derivative. Its most natural definition, therefore, is

$$\left.\frac{d^2y}{dx^2}\right|_n = \frac{\Delta(dy/dx)}{\Delta x} = \frac{\left(dy/dx|_{n+1/2} - dy/dx|_{n-1/2}\right)}{x_{n+1/2} - x_{n-1/2}}, \tag{3.8}$$

[1] A more general form of bisection in which the interval is divided in two *unequal* parts weighted by the value of the function $s_n = [f(s_l)s_u - f(s_u)s_l]/[f(s_l) - f(s_u)]$ has comparable robustness, requires no additional function evaluations per step but a couple of extra multiplications, and converges quicker than plain bisection except in special cases. This is sometimes called the "false-position" root-finding method. It is generally possible to detect automatically when one of those slowly converging special cases is encountered, and to take additional measures.

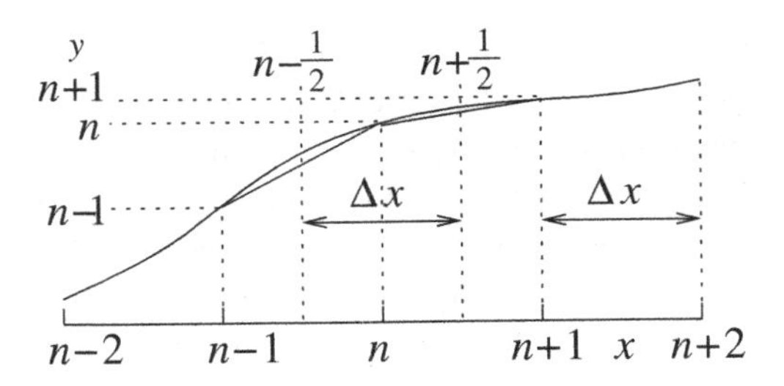

Figure 3.5 Discrete second derivative at n is the difference between the discrete derivatives at $n+1/2$ and $n-1/2$. In a uniform mesh, it is divided by the same Δx.

as illustrated in Fig. 3.5. Because the first derivative is the value at $n+1/2$, the second derivative (the derivative of the first derivative) is the value at a point mid way between $n+1/2$ and $n-1/2$, i.e. at n. Substituting from the previous equation (3.7) we get:[2]

$$\left.\frac{d^2y}{dx^2}\right|_n = \frac{(y_{n+1}-y_n)/\Delta x - (y_n - y_{n-1})/\Delta x}{\Delta x} = \frac{y_{n+1}-2y_n+y_{n-1}}{\Delta x^2}. \quad (3.9)$$

Now think of the entire mesh stretching from $n=1$ to $n=N$. The values y_n at all the nodes can be considered to be a column vector of length N. The second derivative can then be considered to be a matrix operating on that column vector, to give the values of eq. (3.9). So, written in matrix form we have:

$$\begin{pmatrix} d^2y/dx^2|_1 \\ \vdots \\ d^2y/dx^2|_n \\ \vdots \\ d^2y/dx^2|_N \end{pmatrix} = \frac{1}{\Delta x^2}\begin{pmatrix} \ddots & \ddots & 0 & 0 & 0 \\ 1 & -2 & 1 & 0 & 0 \\ 0 & \ddots & \ddots & \ddots & 0 \\ 0 & 0 & 1 & -2 & 1 \\ 0 & 0 & 0 & \ddots & \ddots \end{pmatrix}\begin{pmatrix} y_1 \\ \vdots \\ y_n \\ \vdots \\ y_N \end{pmatrix}, \quad (3.10)$$

where the square $N \times N$ matrix has diagonal elements equal to -2. On the adjacent diagonals, sometimes called subdiagonals (indices $n, n+1$ and $n, n-1$), it has 1; and everywhere else it is zero. This overall form is called tridiagonal.

If we are considering the equation

$$\frac{d^2y}{dx^2} = g(x) \quad (3.11)$$

[2] A shorthand way to remember the result for an order N derivative is that the numerator is the sum of the neighboring y values times the coefficients of a binomial expansion to the power N, and the denominator is Δx^N.

where $g(x)$ is some function (for example $g = -\rho/\epsilon_0$ for our electrostatic example) then the equation is represented by putting the column vector $(d^2y/dx^2|_n)$ equal to the column vector $(g_n) = (g(x_n))$.

3.3.2 Boundary conditions

However, in eq. (3.10), the top left and bottom right corners of the derivative matrix have deliberately been left ambiguous, because that's where the boundary conditions come into play. Assuming they are on the boundaries, the quantities y_1 and y_N are determined not by the differential equation and the function g but by the boundary values. We must adjust the first and last row of the matrix accordingly to represent those boundary conditions. A convenient way to do this when the conditions consist of specifying y_L and y_R at the left and right-hand ends is to write the equation as:

$$\begin{pmatrix} -2 & 0 & 0 & 0 & 0 & 0 & 0 \\ 1 & -2 & 1 & 0 & 0 & 0 & 0 \\ 0 & \ddots & \ddots & \ddots & 0 & 0 & 0 \\ 0 & 0 & 1 & -2 & 1 & 0 & 0 \\ 0 & 0 & 0 & \ddots & \ddots & \ddots & 0 \\ 0 & 0 & 0 & 0 & 1 & -2 & 1 \\ 0 & 0 & 0 & 0 & 0 & 0 & -2 \end{pmatrix} \begin{pmatrix} y_1 \\ y_2 \\ \vdots \\ y_n \\ \vdots \\ y_{N-1} \\ y_N \end{pmatrix} = \begin{pmatrix} -2y_L \\ g_2\Delta x^2 \\ \vdots \\ g_n\Delta x^2 \\ \vdots \\ g_{N-1}\Delta x^2 \\ -2y_R \end{pmatrix}. \tag{3.12}$$

Notice that the first and last rows of the matrix have been made purely diagonal, and the column vector on the right-hand side (call it **h**) uses for the first and last rows the boundary values, and for the others the elements of $\mathbf{g}\Delta x^2$. These adjustments enforce that the first and last values of y are always the boundary values y_L and y_R.[3]

[3] **Enrichment:** An alternative approach to implementing boundary conditions is to regard the boundary positions as outside and not operated upon by the difference matrix. Then its indices run only from 2 to $N-1$, and it has dimensions $(N-2)\times(N-2)$. For Dirichlet conditions, the term in the difference stencils of positions 2 and $N-1$ that contains the boundary value is moved into the right-hand-side source vector. Generically the matrix equation is then

$$\begin{pmatrix} -2 & 1 & 0 & 0 & 0 \\ \ddots & \ddots & \ddots & 0 & 0 \\ 0 & 1 & -2 & 1 & 0 \\ 0 & 0 & \ddots & \ddots & \ddots \\ 0 & 0 & 0 & 1 & -2 \end{pmatrix} \begin{pmatrix} y_2 \\ \vdots \\ y_n \\ \vdots \\ y_{N-1} \end{pmatrix} = \begin{pmatrix} g_2\Delta x^2 - y_L \\ \vdots \\ g_n\Delta x^2 \\ \vdots \\ g_{N-1}\Delta x^2 - y_R \end{pmatrix}.$$

This form keeps the matrix symmetric, which is advantageous for some inversion algorithms.

Once we have constructed this matrix form of the differential equation, which we can write

$$\mathbf{D}\mathbf{y} = \mathbf{h}, \tag{3.13}$$

it is obvious that we can solve it by simply inverting the matrix $\mathbf{D}$ and finding

$$\mathbf{y} = \mathbf{D}^{-1}\mathbf{h}. \tag{3.14}$$

(Or we can use some other appropriate matrix equation solution technique.)

In general, we must make the first and last rows of the matrix equation into discrete expressions of the boundary conditions there. If instead of Dirichlet boundary conditions (value is specified) we are given Neumann conditions, that is, the *derivative* (e.g. $dy/dx|_1$) is specified, a different adjustment of the corners is necessary. The most obvious thing to do is to make the first row of the matrix equation proportional to

$$(-1 \quad 1 \quad 0 \quad \ldots)(\mathbf{y}) = y_2 - y_1 = \Delta x(dy/dx|_1). \tag{3.15}$$

However, this choice does not calculate the derivative at the right place. The expression $(y_2 - y_1)/\Delta x$ is the derivative at $x_{3/2}$ rather than x_1, which is the boundary.[4] So the scheme (3.15) is not properly centered and will give only first-order accuracy.[5] A better extrapolation of the derivative to the boundary is to write instead for the first row

$$\left(-\tfrac{3}{2} \quad 2 \quad -\tfrac{1}{2} \quad 0 \quad \ldots\right)(\mathbf{y}) = -\tfrac{1}{2}(y_3 - y_2) + \tfrac{3}{2}(y_2 - y_1) = \Delta x(dy/dx|_1). \tag{3.16}$$

This is a discrete form of the expression $y'_1 \approx y'_{3/2} - y''_2 . \frac{1}{2}\Delta x$, which is accurate to second order, because it cancels out the first-order error in the derivative. The same treatment applies to a Neumann condition at x_N (but of course using the mirror image of the row given in eq. (3.16)).

If the boundary condition is of a more general form (the so-called Robin condition)

$$Ay + By' + C = 0, \tag{3.17}$$

[4] In some circumstances one can deliberately choose the mesh so as to put the boundary at $x_{3/2}$. Then the implementation represented by eq. (3.15) does put the value at the correct place. This mesh choice is appropriate if the boundary condition is known to be of purely derivative form. It then lends itself to the alternative approach of the previous note, excluding the boundary from $\mathbf{D}$. The top-left coefficient becomes $D_{22} = -1$, and $\Delta x\ (dy/dx|_L)$ is added to the source vector, to implement the boundary condition. Mixed boundary conditions are not handled so easily like this, which is why a second-order accurate form with the boundary at x_1 has been given.

[5] The error in the first derivative y' is approximately $y''_2 . \frac{1}{2}\Delta x$, first order in Δx, but the error in Δy is second order in Δx giving first-order accuracy.

then we want the first row to represent this equation discretely. The natural way to do this, based upon our previous forms, is to make it

$$\left[A\left(1 \quad 0 \quad 0 \quad \ldots\right) + \frac{B}{\Delta x}\left(-\tfrac{3}{2} \quad 2 \quad -\tfrac{1}{2} \quad 0 \quad \ldots\right)\right](\mathbf{y}) = -C. \quad (3.18)$$

In addition to combining the previous inhomogeneous boundary forms this expression can also represent the specification of homogeneous boundary conditions, or in other words logarithmic gradient conditions. When $C = 0$, the boundary condition is $d(\ln y)/dx = y'/y = -A/B$. This form (with $A/B = 1$) is the condition that one might apply to the potential due to a spherically symmetric electrostatic charge at the outer boundary, for example.

It may be preferable in some cases to scale the first row of the matrix equation to make the diagonal entry the same as all the other diagonals, namely -2. This is done by multiplying all the row's elements of $\mathbf{D}$, and the corresponding entry of $\mathbf{h}$ by a factor $-2/D_{11}$, or $-2/D_{NN}$ respectively. This can improve the conditioning of the matrix, making inversion easier and more accurate.

A final type of boundary condition worth discussing is called "periodic." This expression means that the end of the x-domain is considered to be connected to its beginning. Such a situation arises, for example, if the domain is actually a circle in two-dimensional space. But it is also sometimes used to approximate an infinite domain. For periodic boundary conditions it is usually convenient to label the first and last point 0 and N. See Fig. 3.6. They are the same point; so the values at x_0 and x_N are the same. There are then N different points and the discretized differential equation must be satisfied at them all, with the differences wrapping round to the corresponding point across

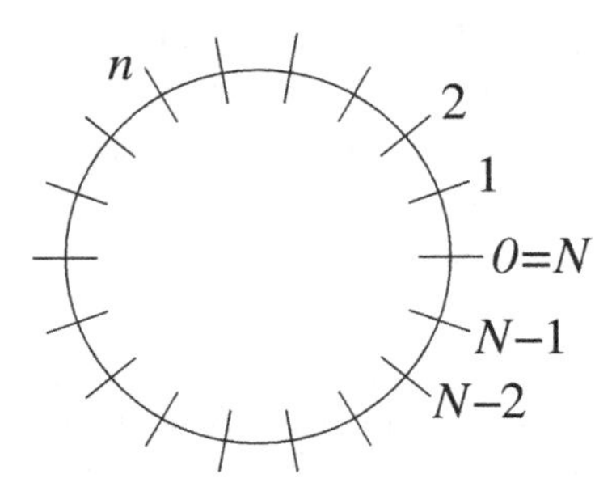

Figure 3.6 Periodic boundary conditions apply when the independent variable is, for example, distance around a periodic domain.

the boundary. The resulting matrix equation is then

$$\begin{pmatrix} -2 & 1 & \dots & 0 & \dots & 0 & 1 \\ 1 & -2 & 1 & 0 & 0 & 0 & 0 \\ \vdots & \ddots & \ddots & \ddots & 0 & 0 & \vdots \\ 0 & 0 & 1 & -2 & 1 & 0 & 0 \\ \vdots & 0 & 0 & \ddots & \ddots & \ddots & \vdots \\ 0 & 0 & 0 & 0 & 1 & -2 & 1 \\ 1 & 0 & \dots & 0 & \dots & 1 & -2 \end{pmatrix} \begin{pmatrix} y_1 \\ y_2 \\ \vdots \\ y_n \\ \vdots \\ y_{N-1} \\ y_N \end{pmatrix} = \begin{pmatrix} g_1 \Delta x^2 \\ g_2 \Delta x^2 \\ \vdots \\ g_n \Delta x^2 \\ \vdots \\ g_{N-1} \Delta x^2 \\ g_N \Delta x^2 \end{pmatrix}, \tag{3.19}$$

which maintains the pattern $1, -2, 1$ for every row, without exception. For the first and last rows, the subdiagonal 1 element that would be outside the matrix is wrapped round to the other end of the row. It gives a new entry in the top-right and bottom-left corners.[6]

3.4 Conservative differences, finite volumes

In our cylindrical fuel-rod example, we had what one might call a "weighted derivative": something more complicated than a Laplacian. One might be tempted to write it in the following way:

$$\frac{d}{dr}\left(r\kappa \frac{dT}{dr}\right) = r\kappa \frac{d^2T}{dr^2} + \frac{d(r\kappa)}{dr}\frac{dT}{dr}, \tag{3.20}$$

and then use the discrete forms for the first and second derivative in this expression. The problem with that approach is that first derivatives are at half mesh points ($n + 1/2$ etc.), while second derivatives are at whole mesh points (n). So it is not clear how best to express this multiple term formula discretely in a consistent manner. In particular, if one adopts an asymmetric form, such as writing $dT/dr|_n \approx (T_{n+1} - T_n)/\Delta x$ (just ignoring the fact that this is really centered at $n + 1/2$, not n), then the error will be of second order in Δx. The scheme will be accurate only to first order. That's bad.

We must avoid that error. But even so, there are various *different* ways to produce schemes that are second-order accurate. Generally the best way is to

[6] With periodic boundary conditions the homogeneous equation ($\frac{d^2y}{dx^2} = 0$) is satisfied by any constant y. An additional condition must therefore be applied to make the solution unique. Moreover, there exists no solution of the differential equation with continuous derivative unless $\int g dx = 0$. These requirements are reflected in the fact that the matrix of this system is singular. If eq. (3.19) is solved by pseudo-inverse, it gives the solution having zero mean: $\sum y_n/N = 0$, using as right-hand side instead of $\Delta x^2 \mathbf{g}$ the quantity $\Delta x^2 (\mathbf{g} - \sum g_n/N)$.

recall that the differential form arose as a *conservation equation.* It was the conservation of energy that required the heat flux through a particular radius cylinder $2\pi r\kappa dT/dr$ to vary with radius only so as to account for the power density at radius r. It is therefore best to develop the second-order differential in this way. First we form dT/dr in the usual discrete form at $n-1/2$ and $n+1/2$. Then we multiply those values by $r\kappa$ *at the same half-mesh positions* $n-1/2$ *and* $n+1/2$. Then we take the difference of those two fluxes, writing:

$$\begin{aligned}\frac{d}{dr}\left(r\kappa\frac{dT}{dr}\right) &= \left(r_{n+1/2}\kappa_{n+1/2}\frac{T_{n+1}-T_n}{\Delta r} - r_{n-1/2}\kappa_{n-1/2}\frac{T_n-T_{n-1}}{\Delta r}\right)\frac{1}{\Delta r}\\ &= \frac{1}{\Delta r^2}[\quad r_{n+1/2}\kappa_{n+1/2}\,T_{n+1}\\ &\qquad -(r_{n+1/2}\kappa_{n+1/2}+r_{n-1/2}\kappa_{n-1/2})\,T_n\\ &\qquad +r_{n-1/2}\kappa_{n-1/2}\,T_{n-1}\,].\end{aligned} \tag{3.21}$$

The big advantage of this expression is that it exactly conserves the heat flux. This property can be seen by considering the exact heat conservation in integral form over the cell consisting of the range $r_{n-1/2} < r < r_{n+1/2}$:

$$2\pi r_{n+1/2}\kappa_{n+1/2}\left.\frac{dT}{dr}\right|_{n+1/2} - 2\pi r_{n-1/2}\kappa_{n-1/2}\left.\frac{dT}{dr}\right|_{n-1/2} = -\int_{r_{n-1/2}}^{r_{n+1/2}} p2\pi r'dr'. \tag{3.22}$$

Then, adding together the versions of this equation for two adjacent positions $n=k,k+1$, the $\left.\frac{dT}{dr}\right|_{k+1/2}$ terms cancel, provided the expression for $r\kappa\frac{dT}{dr}$ is the same at the same n value regardless of which adjacent cell (k or $k+1$) it arises from. This symmetry is present when using $\left.\frac{dT}{dr}\right|_{n+1/2} = (T_{n+1}-T_n)/\Delta r$ and the half-mesh values of $r\kappa$. The sum of the equations is therefore the exact total conservation for the region $r_{k-1/2} < r < r_{k+3/2}$, consisting of the sum of the two adjacent cells. This process can then be extended over the whole domain, proving total heat conservation. Approaching the discrete equations in this way is sometimes called the method of "finite volumes."[7] The finite volume in our illustrative case is the annular region between $r_{n-1/2}$ and $r_{n+1/2}$.

A less satisfactory alternative which remains second-order accurate might be to evaluate the right-hand side of eq. (3.20) using double distance derivatives that are centered at the n mesh as follows

$$\begin{aligned}\frac{d}{dr}\left(r\kappa\frac{dT}{dr}\right) &= \left(r_n\kappa_n\frac{T_{n+1}-2T_n+T_{n-1}}{\Delta r^2} + \frac{r_{n+1}\kappa_{n+1}-r_{n-1}\kappa_{n-1}}{2\Delta r}\frac{T_{n+1}-T_{n-1}}{2\Delta r}\right)\\ &= \frac{1}{\Delta r^2}[\,(r_{n+1}\kappa_{n+1}/4 + r_n\kappa_n - r_{n-1}\kappa_{n-1}/4)\,T_{n+1}\\ &\qquad -2r_n\kappa_n\,T_n\\ &\qquad +(-r_{n+1}\kappa_{n+1}/4 + r_n\kappa_n + r_{n-1}\kappa_{n-1}/4)\,T_{n-1}].\end{aligned} \tag{3.23}$$

[7] On a structured mesh, a finite-volume method is identical to the finite-difference method provided this conservative differencing is used.

None of the coefficients of the Ts in this expression is the same as in eq. (3.21) unless $r\kappa$ is independent of position. This is true even in the case where $r\kappa$ is known only at whole mesh points so the half-point values in eq. (3.21) are obtained by interpolation. Expression (3.23) does not exactly conserve heat flux, which is an important weakness. Expression (3.21) is usually to be preferred.

Worked example. Formulating radial differences

Formulate a matrix scheme to solve by finite differences the equation

$$\frac{d}{dr}\left(r\frac{dy}{dr}\right) + rg(r) = 0 \tag{3.24}$$

with given g and two-point boundary conditions $dy/dr = 0$ at $r = 0$ and $y = 0$ at $r = N\Delta$, on an r grid of uniform spacing Δ.

We write down the finite difference equation at a generic position:

$$\left.\frac{dy}{dr}\right|_{n+1/2} = \frac{y_{n+1} - y_n}{\Delta}.$$

Substituting this into the differential equation, we get

$$\begin{aligned}
-r_n g_n = \frac{d}{dr}\left(r\frac{dy}{dr}\right)_n &= \left(r_{n+1/2}\left.\frac{dy}{dr}\right|_{n+1/2} - r_{n-1/2}\left.\frac{dy}{dr}\right|_{n-1/2}\right)\frac{1}{\Delta} \\
&= \left(r_{n+1/2}\frac{y_{n+1} - y_n}{\Delta} - r_{n-1/2}\frac{y_n - y_{n-1}}{\Delta}\right)\frac{1}{\Delta} \\
&= \left(r_{n+1/2}y_{n+1} - 2r_n y_n + r_{n-1/2}y_{n-1}\right)\frac{1}{\Delta^2}.
\end{aligned} \tag{3.25}$$

It is convenient (and improves matrix conditioning) to divide this equation through by r_n/Δ^2, so that the nth equation reads

$$\left(\frac{r_{n+1/2}}{r_n}\right)y_{n+1} - 2y_n + \left(\frac{r_{n-1/2}}{r_n}\right)y_{n-1} = -\Delta^2 g_n. \tag{3.26}$$

For definiteness we will take the position of the nth grid point to be $r_n = n\Delta$, so n runs from 0 to N. Then the coefficients become $r_{n\pm1/2}/r_n = n \pm 1/2/n = 1 \pm 1/2n$.

The boundary condition at $n = N$ is $y_N = 0$. At $n = 0$ we want $dy/dr = 0$, but we need to use an expression that is centered at $n = 0$, not $n = 1/2$, to give second-order accuracy. Therefore, following eq. (3.16) we write the equation at $n = 0$

$$\Delta\, dy/dx|_0 = -\tfrac{1}{2}(y_2 - y_1) + \tfrac{3}{2}(y_1 - y_0) = \begin{pmatrix} -\tfrac{3}{2} & 2 & -\tfrac{1}{2} & 0 & \ldots \end{pmatrix} (\mathbf{y}) = 0. \tag{3.27}$$

Gathering all our equations into a matrix we have

$$\begin{pmatrix} -\frac{3}{2} & 2 & -\frac{1}{2} & 0 & 0 & 0 & 0 \\ 1-\frac{1}{2} & -2 & 1+\frac{1}{2} & 0 & 0 & 0 & 0 \\ 0 & \ddots & \ddots & \ddots & 0 & 0 & 0 \\ 0 & 0 & 1-\frac{1}{2n} & -2 & 1+\frac{1}{2n} & 0 & 0 \\ 0 & 0 & 0 & \ddots & \ddots & \ddots & 0 \\ 0 & 0 & 0 & 0 & 1-\frac{1}{2(N-1)} & -2 & 1+\frac{1}{2(N-1)} \\ 0 & 0 & 0 & 0 & 0 & 0 & -2 \end{pmatrix} \begin{pmatrix} y_0 \\ y_1 \\ \vdots \\ y_n \\ \vdots \\ y_{N-1} \\ y_N \end{pmatrix} = -\Delta^2 \begin{pmatrix} 0 \\ g_1 \\ \vdots \\ g_n \\ \vdots \\ g_{N-1} \\ 0 \end{pmatrix}. \tag{3.28}$$

Fig. 3.7 shows the solution of an illustrative case.

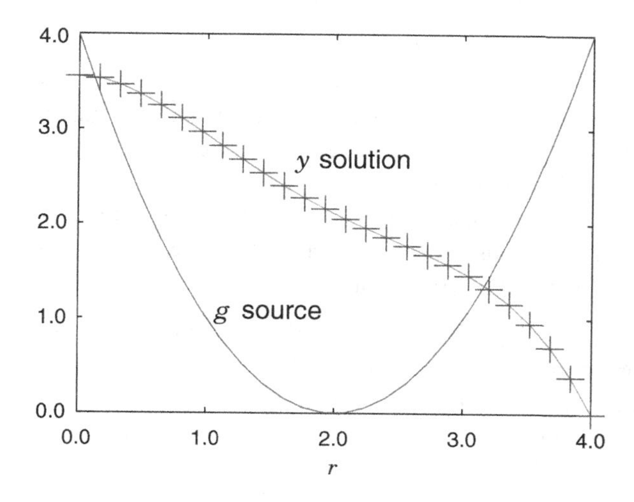

Figure 3.7 Example of the result of a finite-difference solution for y of eq. (3.24) using a matrix of the form of eq. (3.28). The source g is purely illustrative, and is plotted in the figure. The boundary points at the ends of the range of solution are $r = 0$, and $r = N\Delta = 4$. A grid size $N = 25$ is used.

Exercise 3. Solving two-point ordinary differential equations

1. Write a code to solve, using matrix inversion, a two-point ordinary differential equation of the form

$$\frac{d^2y}{dx^2} = f(x)$$

on the x-domain $[0, 1]$, spanned by an equally spaced mesh of N nodes, with Dirichlet boundary conditions $y(0) = y_0$, $y(1) = y_1$.

When you have got it working, obtain your personal expressions for $f(x)$, N, y_0, and y_1 from http://www.essentialnumericalmethods.net/giveassign.html. (Or use $f(x) = a + bx$, $a = 0.15346$, $b = 0.56614$, $N = 44$, $y_0 = 0.53488$, $y_1 = 0.71957$.) And solve the differential equation so posed. Plot the solution.

Submit the following as your solution:

1. Your code in a computer format that is capable of being executed.
2. The expressions of your problem $f(x)$, N, y_0, and y_1.
3. The numeric values of your solution y_j.
4. Your plot.
5. Brief commentary (< 300 words) on what problems you faced and how you solved them.

2. Save your code and make a copy with a new name. Edit the new code so that it solves the ordinary differential equation

$$\frac{d^2y}{dx^2} + k^2y = f(x)$$

on the same domain and with the same boundary conditions, but with the extra parameter k^2. Verify that your new code works correctly for small values of k^2, yielding results close to those of the previous problem.

Investigate what happens to the solution in the vicinity of $k = \pi$.

Describe what the cause of any interesting behavior is.

Submit the following as your solution:

1. Your code in a computer format that is capable of being executed.
2. The expressions of your problem $f(x)$, N, y_0, and y_1.
3. Brief description (< 300 words) of the results of your investigation and your explanation.
4. Back up the explanation with plots if you like.

4
Partial differential equations

4.1 Examples of partial differential equations

Partial differential equations arise in almost every problem that exists in multi-dimensional space. The gradient operator of vector calculus is (in three-dimensional space)

$$\nabla = \left(\frac{\partial}{\partial x}, \frac{\partial}{\partial y}, \frac{\partial}{\partial z}\right), \tag{4.1}$$

a *partial* differential operator.

Partial differential equations also arise when there is time-dependent behavior within a domain that is one-dimensional in space. Then the two dimensions (independent variables) are x and t.

Probably the most important part of computational science and engineering is formulating the calculation in terms of partial differential equations. Once a situation is properly formulated, the numerical techniques we are studying can be applied. But turning a problem in the real world into partial differential equations representing it requires deep knowledge. This section can give only a few examples.

4.1.1 Fluid flow

The flow of a substance that can be considered to be a continuum fluid, such as water, or (collisional) gas, is governed by a hierarchy of equations. They are fundamentally conservation equations of substance (e.g. mass), of momentum, and of energy. The momentum conservation equation is often called the Navier–Stokes equation. It relies on the other equations. Let's derive the conservation of substance or continuity equation.

Consider a substance with density $\rho(\boldsymbol{x})$, which we can consider to be the substance's mass per unit volume (but could alternatively be the number of

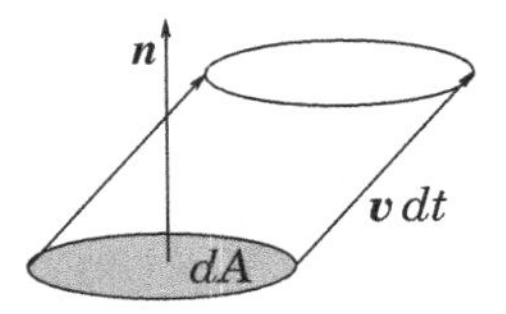

Figure 4.1 The elementary volume of fluid crossing a surface element dA in time element dt is $dt\boldsymbol{v}.\hat{\boldsymbol{n}}dA = dt\boldsymbol{v}.\boldsymbol{dA}$. So the mass per unit time is $\rho\boldsymbol{v}.\boldsymbol{dA}$.

particles per unit volume or electric charge per unit volume). Suppose that this substance has a volumetric source density $S(\boldsymbol{x})$. This quantity denotes the amount (mass) of substance that is created per unit time per unit volume at position $\boldsymbol{x}$. Such creation might be by chemical reaction (e.g. producing the substance CO_2 from reaction of CO and O_2) or it might be by nuclear reaction (e.g. producing Xe^{135} by fission of uranium). The source could also be negative, corresponding to destruction of the substance, for example radioactive disintegration of Xe^{135}, or a balance between creation and destruction. Apart from the processes represented by S, however, the substance is conserved (neither created nor destroyed). If the substance is able to flow, and has a velocity $\boldsymbol{v}(\boldsymbol{x})$, then it can move around and as it does so it gives rise to a flux density of substance $\rho\boldsymbol{v}$ (mass per unit area per unit time). [Flux density represents the amount (mass) of substance per unit time carried by the flow across unit area with normal vector in a certain direction; it is therefore a vector quantity.] Across any small surface (element) $\boldsymbol{dA}$ the flux (mass per unit time) is equal to $\rho\boldsymbol{v}.\boldsymbol{dA}$, as illustrated in Fig. 4.1. Mass conservation is then described by considering some control volume V whose surface is ∂V. For any such volume, the rate of increase of the total amount of substance within the volume must be equal to the total source density within the volume plus the amount flowing in across its surface:

$$\frac{\partial}{\partial t}\int_V \rho d^3x = \int_V S d^3x - \int_{\partial V} \rho\boldsymbol{v}.\boldsymbol{dA}. \tag{4.2}$$

Using Gauss's (divergence) theorem, illustrated in Fig. 4.2, that for any vector field $\boldsymbol{u}$ (here $\boldsymbol{u} = \rho\boldsymbol{v}$), the integral over a closed volume of the divergence is equal to the surface integral of the vector:

$$\int_V \nabla.\boldsymbol{u}\, d^3x = \int_{\partial V} \boldsymbol{u}.\boldsymbol{dA}, \tag{4.3}$$

the surface term can be converted to a volume integral and the mass conservation equation becomes:

$$\int_V \left(\frac{\partial \rho}{\partial t} + \nabla.(\rho\boldsymbol{v}) - S\right) d^3x = 0. \tag{4.4}$$

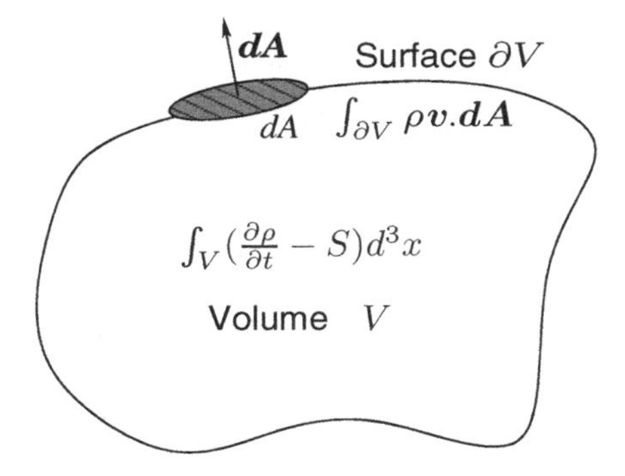

Figure 4.2 Conservation integrals over arbitrary volume V with surface ∂V.

This identity applies for *any* volume V. The only way that can be true is if the *integrand* is everywhere equal to zero:

$$\frac{\partial \rho}{\partial t} + \nabla.(\rho \boldsymbol{v}) - S = 0. \tag{4.5}$$

This is usually called the "continuity" equation. Sometimes, arguably more physically, it is called the "advection" equation. It is a partial differential equation in three space and one time variables.

We'll discuss the rest of the fluid equations later; but for now consider a steady state $\frac{\partial}{\partial t} = 0$, in which the velocity $\boldsymbol{v}(\boldsymbol{x})$ is prescribed everywhere, illustrated in Fig. 4.3. The resulting partial differential equation is

$$\nabla.(\rho \boldsymbol{v}) = \boldsymbol{v}.\nabla \rho + \rho \nabla.\boldsymbol{v} = v_x \frac{\partial \rho}{\partial x} + v_y \frac{\partial \rho}{\partial y} + v_z \frac{\partial \rho}{\partial z} + G(\boldsymbol{x})\rho = S, \tag{4.6}$$

where the components of the velocity v_x, v_y, v_z and the divergence of the velocity $G = \nabla.\boldsymbol{v}$, are prescribed functions of position. From the point of view of solving for ρ as a function of space, this is a linear first-order partial differential equation in three dimensions, with variable (i.e. non-uniform) coefficients v_x, v_y, v_z, and G. It is *inhomogeneous* if $S \neq 0$, but *homogeneous* if $S = 0$ everywhere.

4.1.2 Diffusion

When a substance such as a gas diffuses through some other medium, for example a porous solid, or another (flow free) gas, the flux density $\rho \boldsymbol{v}$ of the substance is frequently proportional to the gradient of its density ρ:

$$\rho \boldsymbol{v} = -D \nabla \rho, \tag{4.7}$$

where D is the diffusion coefficient.

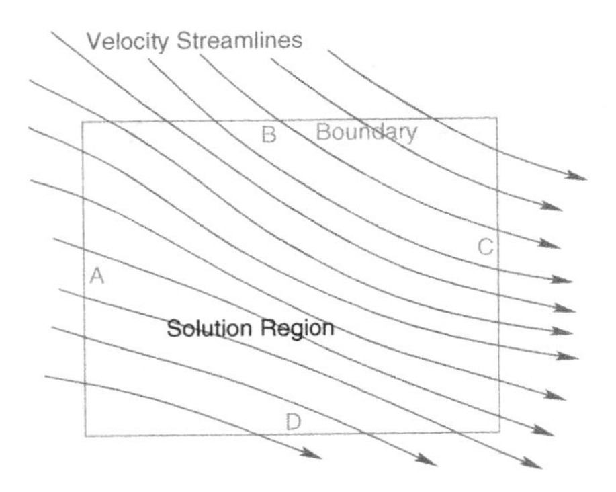

Figure 4.3 The advection equation with prescribed velocity $\boldsymbol{v}$ amounts to integration along the streamlines. Conditions ("initial" conditions) can be applied along the solution region boundaries A and B, but not then along C and D.

For such a diffusion problem, we can eliminate the velocity from the continuity equation (4.5) and obtain the diffusion equation:

$$\frac{\partial \rho}{\partial t} - \nabla.(D\nabla\rho) - S = 0. \tag{4.8}$$

This is a linear partial differential equation that is second order in the spatial derivatives (e.g. if D is uniform the second term becomes Laplacian $-D\nabla^2\rho$); but it is first order in time.

4.1.3 Waves

The equation that governs waves in one dimension, for example small-amplitude compressional vibrations of a column of air in a pipe, or transverse vibrations of a stretched string or a rigid rod, is

$$\frac{\partial^2 \psi}{\partial t^2} = c_s^2 \frac{\partial^2 \psi}{\partial x^2}, \tag{4.9}$$

where c_s is the wave (sound) speed, and ψ represents the wave displacement or perturbed quantity (e.g. pressure). The wave equation is second order in space and time, and linear (provided c_s does not depend on ψ).

4.1.4 Electromagnetism

Maxwell's equations of electromagnetism, relating the electric and magnetic fields $\boldsymbol{E}$ and $\boldsymbol{B}$ to the charge density ρ_q, and the electric current density $\boldsymbol{j}$, in the absence of dielectric or magnetic materials are:

$$\begin{aligned}
\nabla.\boldsymbol{E} &= \rho_q/\epsilon_0 \\
\nabla \times \boldsymbol{E} &= -\frac{\partial \boldsymbol{B}}{\partial t} \\
\nabla.\boldsymbol{B} &= 0 \\
\nabla \times \boldsymbol{B} &= \mu_0 \boldsymbol{j} + \mu_0 \epsilon_0 \frac{\partial \boldsymbol{E}}{\partial t}
\end{aligned} \tag{4.10}$$

where $\epsilon_0 = 8.85 \times 10^{-12}$ F/m is the permittivity of free space and $\mu_0 = 4\pi \times 10^{-7}$ H/m is the permeability of free space. These fundamental constants satisfy $\mu_0 \epsilon_0 = 1/c^2$, where c is the speed of light.

These are a system of partial differential equations. There appear to be eight in all because the curl equations are vector equations (three equations in one) and the divergence equations are single equations. However, there's some built-in redundancy in the equations that reduces their effective number to six (equal to the number of dependent variables in the components of the electric and magnetic fields).

Only rarely does one solve the full set of equations numerically. More usually one is interested in simplified special cases. For example, if time dependence can be ignored, then $\nabla \times \boldsymbol{E} = 0$, which is a sufficient condition to allow the electric field to be expressed as the gradient of a scalar potential $\boldsymbol{E} = -\nabla\phi$. In that case, the potential satisfies

$$-\nabla.\boldsymbol{E} = -\nabla.(-\nabla\phi) = \nabla^2\phi = \frac{\partial^2\phi}{\partial x^2} + \frac{\partial^2\phi}{\partial y^2} + \frac{\partial^2\phi}{\partial z^2} = -\rho_q/\epsilon_0, \tag{4.11}$$

which is Poisson's equation[1]. Poisson's equation sets a *second-order* differential, the *Laplacian* (∇^2) of the potential equal to a (presumably prescribed) function $-\rho_q/\epsilon_0$.

4.2 Classification of partial differential equations

In the theory of partial differential equations, there are three types of differential equation. They are called: hyperbolic, parabolic, and elliptic. To understand this classification rigorously would take us far beyond our scope, but it is important for computation because the methods of solving the different types of equation are different.

[1] The subscript q on ρ_q reminds us that this is charge density, not mass density ρ here.

For our purposes, the classification can be considered as follows for second-order equations. Write the general linear second-order partial differential equation governing dependent variable ψ for independent variables x_i as

$$\sum_{i,j} c_{ij}\frac{\partial}{\partial x_i}\frac{\partial}{\partial x_j}\psi + \sum_i c_i\frac{\partial}{\partial x_i}\psi + c\psi = const. \tag{4.12}$$

A specific example might be

$$A\frac{\partial^2}{\partial x^2}\psi + 2B\frac{\partial^2}{\partial x \partial y}\psi + C\frac{\partial^2}{\partial y^2}\psi = 0. \tag{4.13}$$

Consider the coefficients c_{ij}, and regard them as defining a surface in the multidimensional space via the quadratic form associated with them:

$$\sum_{i,j} c_{ij}x_i x_j + \sum_i c_i x_i = const. \tag{4.14}$$

Specifically:

$$Ax^2 + 2Bxy + Cy^2 = const. \tag{4.15}$$

Then the partial differential equation is hyperbolic, parabolic, or elliptic according to whether the surface so defined is itself hyperbolic, parabolic, or elliptic.

For our two-dimensional specific example, the surface is

hyperbolic if $B^2 - AC > 0$; for example $B = 0$, $C = -A$, $x^2 - y^2 = const.$
parabolic if $B^2 - AC = 0$; for example $B = C = 0$, $x^2 - y = const.$
elliptic if $B^2 - AC < 0$; for example $B = 0$, $C = A$, $x^2 + y^2 = const.$

And so is the equation. The examples we have given previously illustrate characteristically these different types.

The *wave equation* is *hyperbolic*.
The *diffusion equation* is *parabolic*.
The *Poisson equation* is *elliptic*.

The first-order equation for fluid flow with prescribed velocity (advection equation) is ambiguous under this classification because the surface for a first-order form is a plane. But a plane can be considered a degenerate hyperbola, because it extends to infinity (unlike an ellipse, which is finite in extent). The first-order system for one scalar dependent variable is always hyperbolic.[2]

Loosely speaking, second-order hyperbolic equations are wave-like, and elliptic equations are steady-flux-conservation-like. A hyperbolic or parabolic

[2] Vector dependent-variable problems are hyperbolic if the matrix of the coefficients of their differentials is diagonizable with real eigenvalues, as we'll see later.

equation generally has at least one independent variable that is like time, as well as others that are like space. An elliptic problem is more like a steady (non-time-varying) problem in multiple space dimensions. The classification is more fundamentally to do with whether a problem of order N with N boundary conditions on a single surface (the multidimensional equivalent of an initial-value problem) can be solved or not. Such a problem is called a Cauchy problem. Generally, hyperbolic equations are those for which the Cauchy problem can be solved; for elliptic equations it cannot.

The most important distinction from the viewpoint of numerical solution of partial differential equations is with respect to the boundary conditions. Are these all applied on a single open surface, such as the plane at $t = 0$? If so (hyperbolic and parabolic equations only) then this is an *initial-value problem*, where we are trying to solve for the time evolution and we can integrate forward in time from the initial conditions. For example, recalling Fig. 4.3 for the (hyperbolic) advection equation, it can immediately be seen that because the equation governs the variation of density along a streamline, it could make no sense to apply conditions at two different points along the one streamline. Alternatively, are the conditions applied on the boundary of a closed domain (as illustrated in Fig. 4.4)? If so (elliptic equations) then the solution everywhere depends upon all the boundary information, it is called a *boundary-value problem*. Then we generally can't successfully integrate

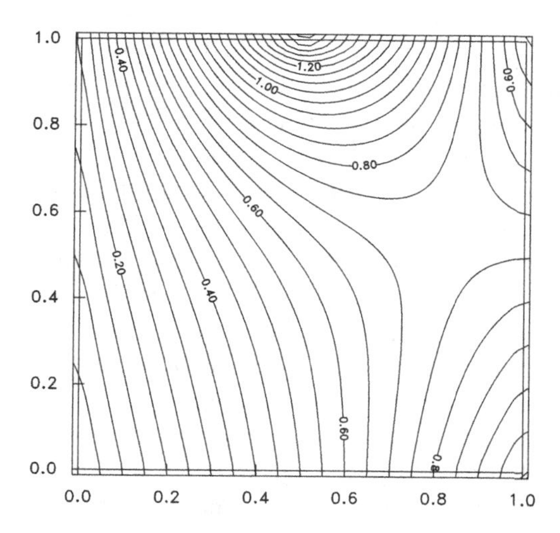

Figure 4.4 An elliptic equation, such as Laplace's equation, has boundary conditions applied round an entire closed surface. In this case, the potential is prescribed everywhere on the square domain boundary, and contours of potential are plotted.

from the boundary. "Shooting" doesn't work in multiple dimensions; so we can't turn the problem into an interative initial-value problem the way we did in one dimension. We have to converge to the full solution everywhere simultaneously.

4.3 Finite-difference partial derivatives

As with ordinary differential equations, a key step in implementing numerical solution of partial differential equations is the expression of the derivatives in terms of the discrete numerical representation of the variables. There are many possible such discrete representations. One might be in terms of the Fourier coefficients a_{ij} of a discrete Fourier representation such as:

$$\psi = \sum_{i,j} a_{ij} \sin(\pi ix/L_x) \sin(\pi jy/L_y) \tag{4.16}$$

(where we've chosen to rule out cosine terms by presuming the boundary conditions to be $\psi = 0$ at $x = 0, L_x$ or $y = 0, L_y$). There are some equations in which it is computationally efficient to work in terms of such coefficients, especially in coordinate directions that are ignorable. Far more often, though, the representation of our variables is on a discrete spatial mesh with finite spacing in the independent variables.

A mesh is considered to be "structured" if the points at which the solution is to be found are in a regular order in each of the coordinate directions. The most obvious choice is cartesian coordinates. The mesh nodes would then be at positions (x_n, y_m, z_l), where each array x_n, y_m, etc., is ordered $x_{n-1} < x_n < x_{n+1}$, etc. The mesh is then "uniform" if the spacing between nodes is the same for all nodes: $x_{n+1} - x_n = \Delta x$ for all n, and similarly for y, z. Structured meshes are also possible on general curvilinear coordinates, for example cylindrical or spherical coordinates, provided the domain remains "rectangular" in the coordinates; in other words, provided for some set of coordinates (ξ, η, ζ), the ξ_n are independent of η_m and ζ_l, and so on. Fig. 4.5 gives two examples. For such structured meshes, the most natural and adaptable finite representation is by finite differences. By contrast, an *unstructured* grid can have any degree of connectedness. Its cells will not generally be quadrilaterals (in two dimensions) or hexahedrons (in three dimensions) and it is far less clear how to construct a finite-difference scheme. Fig. 4.6 shows an example of a triangular unstructured mesh.[3] We'll stick to structured meshes here.

[3] Constructed with the DistMesh routines from http://persson.berkeley.edu/distmesh/.

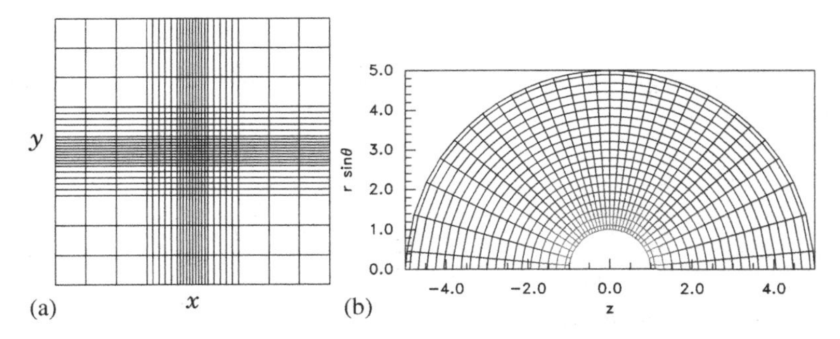

Figure 4.5 Structured grids. (a) Rectangular, but non-uniform. (b) Curvilinear, but still structured.

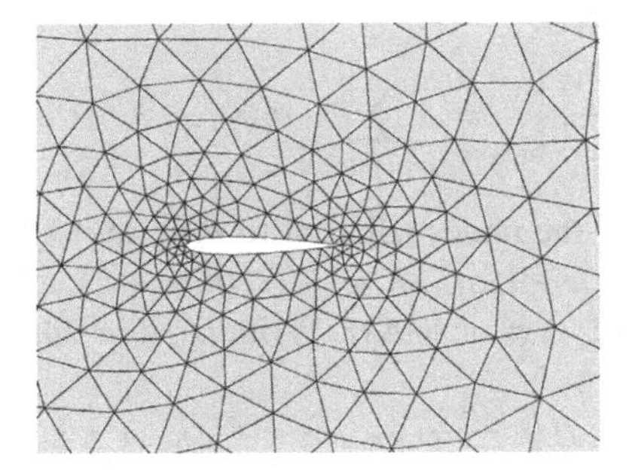

Figure 4.6 A region of an unstructured mesh. Such meshes are designed to give greater resolution in areas of more rapid variation, often close to boundaries of complicated or subtle shapes. This example uses a triangular mesh but other choices are possible.

Just as in one dimension we use a representation

$$\left.\frac{d\psi}{dx}\right|_{n+1/2} = \frac{\psi_{n+1} - \psi_n}{x_{n+1} - x_n}, \tag{4.17}$$

so in two dimensions we can generalize this to partial derivatives of a dependent variable $\psi(x, y)$ in x and y on mesh positions m, n as

$$\left.\frac{\partial\psi}{\partial x}\right|_{n+1/2,m} = \frac{\psi_{n+1,m} - \psi_{n,m}}{x_{n+1} - x_n}, \qquad \left.\frac{\partial\psi}{\partial y}\right|_{n,m+1/2} = \frac{\psi_{n,m+1} - \psi_{n,m}}{y_{m+1} - y_m}, \tag{4.18}$$

with obvious generalization to higher dimensions. Two (or more) indices denote the mesh node (or half-node position) where the value of the variable is to be considered evaluated.

Higher derivatives are formed by taking appropriate differences of derivative expressions. So

$$\left.\frac{\partial^2\psi}{\partial x^2}\right|_{n,m} = \frac{\left(\left.\frac{\partial\psi}{\partial x}\right|_{n+1/2,m} - \left.\frac{\partial\psi}{\partial x}\right|_{n-1/2,m}\right)}{x_{n+1/2} - x_{n-1/2}} \tag{4.19}$$

and

$$\left.\frac{\partial^2\psi}{\partial y^2}\right|_{n,m} = \frac{\left(\left.\frac{\partial\psi}{\partial y}\right|_{n,m+1/2} - \left.\frac{\partial\psi}{\partial y}\right|_{n,m-1/2}\right)}{y_{m+1/2} - y_{m-1/2}}, \tag{4.20}$$

into which we substitute the expressions from eq. (4.18). That leads to a sum of coefficients times the values of ψ at three mesh points.

If we are dealing, for example, with Poisson's equation, $\nabla^2\psi = \rho$, then the entire finite difference expression becomes:

$$\left.\frac{\partial^2\psi}{\partial x^2}\right|_{n,m} + \left.\frac{\partial^2\psi}{\partial y^2}\right|_{n,m} = \sum_{i=adjacent} a_i(\psi_i - \psi_{n,m}) = \rho_{n,m}, \tag{4.21}$$

where the sum is over the nodes adjacent to n, m; in other words, i takes the four cases $(n-1, m)$, $(n+1, m)$, $(n, m-1)$, $(n, m+1)$, and for uniform mesh the coefficients a_i are $1/\Delta x^2$ or $1/\Delta y^2$. Written out in full

$$\frac{1}{\Delta x^2}(\psi_{n+1,m} + \psi_{n-1,m}) + \frac{1}{\Delta y^2}(\psi_{n,m+1} + \psi_{n,m-1}) - \left(\frac{2}{\Delta x^2} + \frac{2}{\Delta y^2}\right)\psi_{n,m} = \rho_{n,m}. \tag{4.22}$$

This form naturally generalizes immediately to higher dimensions. It has a standard structure represented by eq. (4.21), namely that the second-order differential operator is represented by the sum over all the adjacent nodes of coefficients times the adjacent values, minus the sum of all the coefficients times the central value. This sum is called a "stencil," representing the differential. The sum of all its coefficients (including the coefficient of the central value) is zero, because if ψ is uniform, $\nabla^2\psi$ is zero. Written out geometrically for two dimensions the coefficients form a cross pattern, which for the uniform-mesh Laplacian is:

$$\begin{array}{r|ccc} m+1 & . & 1/\Delta y^2 & . \\ m & 1/\Delta x^2 & -(2/\Delta x^2 + 2/\Delta y^2) & 1/\Delta x^2 \\ m-1 & . & 1/\Delta y^2 & . \\ \hline & n-1 & n & n+1 \end{array} . \tag{4.23}$$

Other linear second-order differential operators, or non-uniform or curvilinear meshes, will have different coefficients, but will still have the same geometric shape, and will still have the central coefficient equal to minus the sum of the others.

The stencil represented by eq. (4.23) is called a "star stencil." It includes only the adjacent point along coordinate directions. It is second-order accurate. One can make differential operator approximations whose errors are of higher-order accuracy. Such modified stencils might fill in the corners of the 3×3 matrix or even go additional steps beyond the edges, using appropriate coefficients. Although there may be specific applications in which such expanded stencils are appropriate, they are rather rare. The main focus is generally on ensuring that the coefficients are calculated well centered so as to maintain the second-order accuracy. This is not automatic in non-uniform mesh cases.

Worked example. Cylindrical differences

Determine the type classification of the following partial differential equation, governing a problem in cylindrical coordinates r, θ:

$$\nabla^2 \psi \equiv \frac{1}{r}\frac{\partial}{\partial r}\left(r\frac{\partial \psi}{\partial r}\right) + \frac{1}{r^2}\frac{\partial^2 \psi}{\partial \theta^2} = -k^2 \psi. \qquad (4.24)$$

Obtain an appropriate difference stencil representing the differential operator, at grid point r_n, θ_m for a grid with uniform mesh in r and θ such that $r_n = n\Delta r$, and $\theta_m = m\Delta\theta$.

The quadratic form in x and y arising from the coefficients for the derivatives with respect to r and θ is

$$x^2 + \frac{1}{r^2}y^2 + \frac{1}{r}x = const. \qquad (4.25)$$

Since r^2 is always positive, this is the equation of an ellipse (the ratio of its axes is $1/r$). The differential equation is *elliptic*. We can apply a condition around an entire closed boundary in the r–θ plane. In practice θ is periodic, so there usually isn't a true boundary at $\theta = 0$. At $r = 0$ (if the domain extends to it), $\partial\psi/\partial r = 0$. The boundary conditions are applied at fixed r positions.

To obtain the stencil, start by writing down the first-order partial derivatives in the coordinate directions:

$$\left.\frac{\partial \psi}{\partial r}\right|_{n+1/2,m} = \frac{\psi_{n+1,m} - \psi_{n,m}}{r_{n+1} - r_n}, \qquad \left.\frac{\partial \psi}{r\partial \theta}\right|_{n,m+1/2} = \frac{\psi_{n,m+1} - \psi_{n,m}}{r_n(\theta_{m+1} - \theta_m)}. \qquad (4.26)$$

Then substitute into the second-derivative forms:

$$\begin{aligned}\frac{1}{r}\frac{\partial}{\partial r}\left(r\frac{\partial\psi}{\partial r}\right)_{n,m} &= \frac{1}{r_n}\left(r_{n+1/2}\frac{\psi_{n+1,m}-\psi_{n,m}}{r_{n+1}-r_n} - r_{n-1/2}\frac{\psi_{n,m}-\psi_{n-1,m}}{r_n-r_{n-1}}\right)\frac{1}{r_{n+1/2}-r_{n-1/2}} \\ &= \left[\left(1+\frac{1}{2n}\right)(\psi_{n+1,m}-\psi_{n,m}) - \left(1-\frac{1}{2n}\right)(\psi_{n,m}-\psi_{n-1,m})\right]\frac{1}{\Delta r^2} \\ &= \left[\left(1+\frac{1}{2n}\right)\psi_{n+1,m} - 2\psi_{n,m} + \left(1-\frac{1}{2n}\right)\psi_{n-1,m}\right]\frac{1}{\Delta r^2} \quad (4.27)\end{aligned}$$

and

$$\begin{aligned}\left.\frac{1}{r^2}\frac{\partial^2\psi}{\partial\theta^2}\right|_{n,m} &= \frac{1}{r_n^2}\left(\frac{\psi_{n,m+1}-\psi_{n,m}}{\theta_{m+1}-\theta_m} - \frac{\psi_{n,m}-\psi_{n,m-1}}{\theta_m-\theta_{m-1}}\right)\frac{1}{\theta_{m+1/2}-\theta_{m-1/2}} \\ &= \left(\psi_{n,m+1} - 2\psi_{n,m} + \psi_{n,m-1}\right)\frac{1}{r_n^2\Delta\theta^2}. \quad (4.28)\end{aligned}$$

Then we see that the required stencil for the differential operator is

$m+1$	$\cdot$	$1/(r_n\Delta\theta)^2$	$\cdot$
m	$\left(1-\frac{1}{2n}\right)/\Delta r^2$	$-2\left[1/\Delta r^2 + 1/(r_n\Delta\theta)^2\right]$	$\left(1+\frac{1}{2n}\right)/\Delta r^2$.
$m-1$	$\cdot$	$1/(r_n\Delta\theta)^2$	$\cdot$
	$n-1$	n	$n+1$

(4.29)

Exercise 4. Partial differential equations

1. Determine whether the following partial differential equations, in which p and q are arbitrary real constants, are elliptic, parabolic, or hyperbolic:

(a) $p^2\frac{\partial^2\psi}{\partial x^2} + q^2\frac{\partial^2\psi}{\partial y^2} = 0$,

(b) $p^2\frac{\partial^2\psi}{\partial x^2} - q^2\frac{\partial^2\psi}{\partial y^2} = \psi$,

(c) $\frac{\partial^2\psi}{\partial x^2} + 4\frac{\partial^2\psi}{\partial x\partial y} + \frac{\partial^2\psi}{\partial y^2} = 0$,

(d) $\frac{\partial^2\psi}{\partial x^2} + 2\frac{\partial^2\psi}{\partial x\partial y} + \frac{\partial^2\psi}{\partial y^2} = 0$,

(e) $\frac{\partial^2\psi}{\partial x^2} + p\frac{\partial\psi}{\partial y} = \psi$,

(f) $\frac{\partial^2\psi}{\partial x^2} + \frac{\partial^2\psi}{\partial y^2} + \frac{\partial^2\psi}{\partial z^2} = 0$, and

(g) $p\frac{\partial\psi}{\partial x} + qy\frac{\partial\psi}{\partial y} = 1$.

2. Write a computer code function[4] to evaluate the difference stencil in two dimensions for the anisotropic partial differential operator, $\mathcal{L} = \frac{\partial^2}{\partial x^2} + 2\frac{\partial^2}{\partial y^2}$.

[4] For object-oriented purists, this could be a "method."

The code function is to operate on a quantity $f(x, y) = f_{ij}$, represented as a matrix of the values at discrete points on a structured, equally-spaced, two-dimensional mesh with N_x and N_y nodes in the x and y directions, spanning the intervals $0 \le x \le L_x$, $0 \le y \le L_y$. The function should accept parameters $N_x, N_y, L_x, L_y, i, j, f$ and return the corresponding finite-difference expression for $g_{ij} = \mathcal{L}f$ at mesh point i, j.

Write also a test program to construct $f(x, y) = (x^2 + y^2/2)$ on the mesh nodes, giving f_{ij}, and call your stencil function, with f and the corresponding N_x, N_y, L_x, L_y as arguments, to evaluate g_{ij} and print it.

Submit the following as your solution:

1. Your code in a computer format that is capable of being executed, citing the language it is written in.
2. A brief answer to the following: Will your function work at the boundaries, $x = 0, L_x$, or $y = 0, L_y$? If not, what is needed to make it work there?
3. The values of g_{ij} for four different nodes corresponding to two different interior i and two different interior j, when $N_x = N_y = 10$, $L_x = L_y = 10$.
4. Brief answer to: Are there inefficiencies in using a code like this to evaluate $\mathcal{L}f$ everywhere on the mesh? If so, how might those inefficiencies be avoided?

5

Diffusion. Parabolic partial differential equations

5.1 Diffusion equations

The diffusion equation,

$$\frac{\partial \psi}{\partial t} = \nabla.(D\nabla\psi) + s, \tag{5.1}$$

arises in heat conduction, neutron transport, particle diffusion, and numerous other situations. There is a clear difference between the time variable t and the spatial variables $\boldsymbol{x}$. We'll talk mostly for brevity as if there is only one spatial dimension x, but this discussion can readily be generalized to multiple spatial dimensions. The highest time derivative is $\partial/\partial t$, first order. The highest spatial derivative is second order $\partial^2/\partial x^2$. The equation is classified as parabolic. Consequently, boundary conditions are not applied all around a closed contour (in the x–t plane) but generally only at the two ends of the spatial range, and at one "initial" time, as illustrated in Fig. 5.1. The dependent variable solution is propagated from the initial condition forward in time (conventionally drawn as the vertical direction).

5.2 Time-advance choices and stability

5.2.1 Forward time, centered space

For simplicity we'll take D to be uniform in one cartesian dimension, and the meshes to be uniform with spacing Δx and Δt. Then one way to write the equation in discrete finite differences is

$$\frac{\psi_j^{(n+1)} - \psi_j^{(n)}}{\Delta t} = D\frac{\psi_{j+1}^{(n)} - 2\psi_j^{(n)} + \psi_{j-1}^{(n)}}{\Delta x^2} + s_j^{(n)}. \tag{5.2}$$

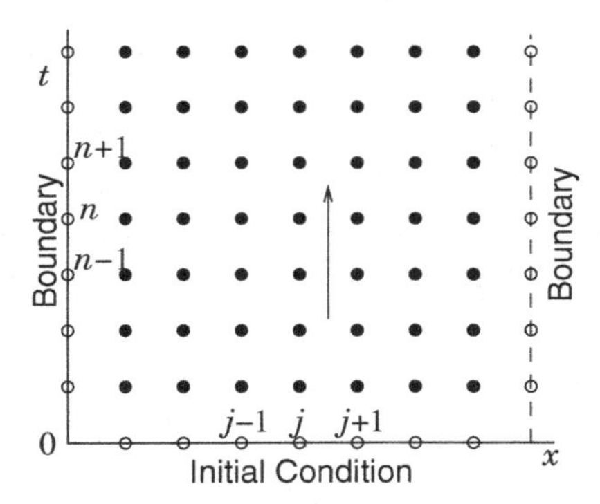

Figure 5.1 Solving a diffusion problem generally requires a combination of spatial boundary conditions and temporal initial condition. Then the solution is propagated upward (forward in time), to fill in the multidimensional (time and space) solution domain.

We use (n) to denote time index, and put it as a superscript to distinguish it from the space indices j $[k, l]$. Of course this notation is *not* raising to a power. Notice in this equation the *second-order* derivative in space is naturally centered and symmetric. However, the time derivative is not centered in time. It is really the value at $n + 1/2$, not at the time index of everything else: n. This scheme is therefore *F*orward in *T*ime, but *C*entered in *S*pace (FTCS); see Fig. 5.2. We immediately know from our previous experience that, because it is not centered in time, this scheme's accuracy is going to be only first order in Δt. Also, this scheme is *explicit* in time. The ψ at $n+1$ is obtained using only prior (n) values of the other quantities:

$$\psi_j^{(n+1)} = \psi_j^{(n)} + \frac{D\Delta t}{\Delta x^2}(\psi_{j+1}^{(n)} - 2\psi_j^{(n)} + \psi_{j-1}^{(n)}) + \Delta t\, s_j^{(n)}. \tag{5.3}$$

A question then arises as to whether this scheme is *stable*. For an ordinary differential equation, we saw that with explicit integration there was a maximum step size that could be allowed before the scheme became unstable. The same is true for hyperbolic and parabolic partial differential equations.

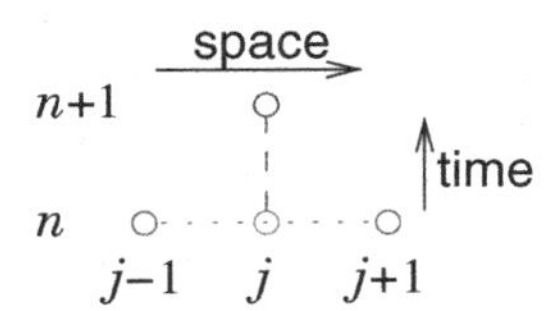

Figure 5.2 Forward time, centered space (FTCS) difference scheme.

For stability analysis, we ignore the source S (because we are really analysing the deviation of the solution[1]). However, even so, it's a bit difficult to see immediately how to evaluate the amplification factor, because for partial differential equations there is variation in the spatial dimension(s) that has to be accounted for. It wasn't present for ordinary differential equations. The way this is generally handled is to turn the partial differential equation into an ordinary differential equation by examining separately all the *Fourier components* of the spatial variation. This sort of analysis is called Von Neumann stability analysis. It gives a precisely correct answer only for uniform grids and coefficients, but it is usually approximately correct, and hence in practice very useful even for non-uniform cases.

A Fourier component varies in space like $\exp(ik_x x)$ where k_x is the wave number in the x-direction (and i is here the square root of minus 1). For such a Fourier component, $\psi_j \propto \exp(ik_x \Delta x j)$, so that $\psi_{j+1} = \exp(ik_x \Delta x)\psi_j$ and $\psi_{j-1} = \exp(-ik_x \Delta x)\psi_j$. Therefore,

$$\psi_{j+1} - 2\psi_j + \psi_{j-1} = (\mathrm{e}^{ik_x \Delta x} - 2 + \mathrm{e}^{-ik_x \Delta x})\psi_j = -4\sin^2(\frac{k_x \Delta x}{2})\psi_j. \quad (5.4)$$

Then, substituting into eq. (5.3) for this Fourier component, we find

$$\psi^{(n+1)} = \underbrace{\left[1 - \frac{D\Delta t}{\Delta x^2} 4\sin^2(\frac{k_x \Delta x}{2})\right]}_{\text{Amplification factor}} \psi^{(n)}. \quad (5.5)$$

The amplification factor from each step to the next is the square-bracket term. If it has a magnitude greater than 1, then instability will occur. If D is negative it will in fact be greater than 1. This instability is not a numerical instability, though. It is a *physical* instability. The diffusion coefficient must be positive otherwise the diffusion equation is unstable regardless of numerical methods. So D must be positive; and so are Δt, Δx. Therefore, numerical instability will arise if the magnitude of the second (negative) term in the amplification factor exceeds 2.

If $k_x \Delta x$ is small, then that will make the second term small and unproblematic. We are most concerned about larger k_x values that can make $\sin^2(k_x \Delta x/2)$ approximately unity. In fact, the largest k_x value that can be represented on a finite grid[2] is such that the phase difference ($k_x \Delta x$) between adjacent values is π radians. That corresponds to a solution that oscillates in sign between adjacent nodes. For that Fourier component, therefore, $\sin^2(k_x \Delta x/2) = 1$.

[1] In effect, we linearize about an exact solution of the equations and the ψ we then analyse is the (presumed small) difference between the solution we obtain and the exact solution. In the text we avoid encumbrance of the notation by not drawing explicit attention to this fact.

[2] The Nyquist limit.

Stability requires *all* Fourier modes to be stable, including the worst mode that has $\sin^2(k_x\Delta x/2) = 1$. Therefore, the condition for stability is

$$\frac{4D\Delta t}{\Delta x^2} < 2. \tag{5.6}$$

There is, for the FTCS scheme, a maximum stable timestep equal to $\Delta x^2/2D$.

Incidentally, the fact that Δt must therefore be no bigger than something proportional to Δx^2 makes the first-order accuracy in time less of a problem. In fact, for a timestep at the stability limit, as we decrease Δx, improving the spatial accuracy proportional to Δx^2 because of the second-order accuracy in space, we also improve the temporal accuracy by the same factor, proportional to Δx^2 because $\Delta t \propto \Delta x^2$.

5.2.2 Backward time, centered space. Implicit scheme

In order to counteract the instability, we learned earlier that *implicit* schemes are helpful. The natural implicit time advance is simply to say that we use the values of the updated variables to evaluate the rest of the equation, instead of the prior values:

$$\psi_j^{(n+1)} = \psi_j^{(n)} + \frac{D\Delta t}{\Delta x^2}(\psi_{j+1}^{(n+1)} - 2\psi_j^{(n+1)} + \psi_{j-1}^{(n+1)}) + \Delta t\, s_j^{(n+1)}. \tag{5.7}$$

This is a Backward in Time, but Centered in Space (BTCS) scheme, illustrated in Fig. 5.3. We'll see a little later how to actually solve this equation for the values at $n+1$, but we can do the same stability analysis on it without knowing. The combination for the spatial Fourier mode is just as in eq. (5.4), so the update equation (ignoring S) for a Fourier mode is

$$\underbrace{\left[1 + \frac{D\Delta t}{\Delta x^2} 4\sin^2(\frac{k_x\Delta x}{2})\right]}_{\text{Inverse of amplification factor}} \psi^{(n+1)} = \psi^{(n)}. \tag{5.8}$$

The amplification factor is the inverse of the square-bracket factor on the left. That square bracket has magnitude always greater than one. Therefore,

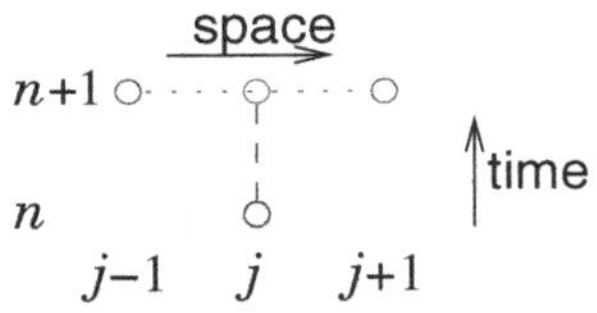

Figure 5.3 Backward time, centered space (BTCS) difference scheme.

the BTCS scheme is *unconditionally stable*. We can take timesteps as large as we like.

It turns out, however, that the only first-order accuracy of this scheme, like the FTCS scheme, means that we don't generally want much to exceed the previously calculated stability limit. If we do so, we increasingly sacrifice temporal accuracy, even though not stability.

5.2.3 Partially implicit, Crank–Nicholson schemes

The best choice, for optimizing the efficiency of the numerical solution of diffusive problems is to use a scheme that is part forward and part backward. A combination of forward and backward, in which θ is the weight of the implicit or backward proportion, is to write

$$\psi_j^{(n+1)} = \psi_j^{(n)} + \theta \left[\frac{D\Delta t}{\Delta x^2}(\psi_{j+1}^{(n+1)} - 2\psi_j^{(n+1)} + \psi_{j-1}^{(n+1)}) + \Delta t\, s_j^{(n+1)} \right]$$
$$+ (1-\theta) \left[\frac{D\Delta t}{\Delta x^2}(\psi_{j+1}^{(n)} - 2\psi_j^{(n)} + \psi_{j-1}^{(n)}) + \Delta t\, s_j^{(n)} \right]. \quad (5.9)$$

This is sometimes called the "θ-implicit" scheme. The amplification factor is straightforwardly

$$A = \frac{1 - (1-\theta)\frac{4D\Delta t}{\Delta x^2}\sin^2(\frac{k_x \Delta x}{2})}{1 + \theta \frac{4D\Delta t}{\Delta x^2}\sin^2(\frac{k_x \Delta x}{2})}. \quad (5.10)$$

If $\theta \geq 1/2$ then $|A| \leq 1$, and the scheme is always stable. If $\theta < 1/2$, then $|A| \leq 1$ requires the stability criterion

$$\Delta t < \frac{\Delta x^2}{2D(1-2\theta)}. \quad (5.11)$$

Thus the minimum degree of implicitness that guarantees stability for all sizes of timestep is $\theta = 1/2$. This choice is called the "Crank–Nicholson" scheme.

It has a major advantage beyond stability. It is centered in time. That means it is second-order accurate in time (as well as space). This accuracy makes it useful to take bigger steps than would be allowed by the (explicit advance) stability limit.

5.3 Implicit advancing matrix method

An implicit or partially implicit scheme for advancing a parabolic equation generally results in equations containing more than one spatial point value at the updated time, for example $\psi_j^{(n+1)}$, $\psi_{j-1}^{(n+1)}$, $\psi_{j+1}^{(n+1)}$. Such an equation for all spatial positions can be written as a matrix equation. Gathering together the terms at n and at $n+1$ from eq. (5.9) it can be written

$$\begin{pmatrix} \ddots & \ddots & 0 & 0 & 0 \\ \ddots & \ddots & \ddots & 0 & 0 \\ 0 & B_- & B_0 & B_+ & 0 \\ 0 & 0 & \ddots & \ddots & \ddots \\ 0 & 0 & 0 & \ddots & \ddots \end{pmatrix} \begin{pmatrix} \psi_1^{(n+1)} \\ \vdots \\ \psi_j^{(n+1)} \\ \vdots \\ \psi_J^{(n+1)} \end{pmatrix} = \begin{pmatrix} \ddots & \ddots & 0 & 0 & 0 \\ \ddots & \ddots & \ddots & 0 & 0 \\ 0 & C_- & C_0 & C_+ & 0 \\ 0 & 0 & \ddots & \ddots & \ddots \\ 0 & 0 & 0 & \ddots & \ddots \end{pmatrix} \begin{pmatrix} \psi_1^{(n)} \\ \vdots \\ \psi_j^{(n)} \\ \vdots \\ \psi_J^{(n)} \end{pmatrix} + \begin{pmatrix} s_1 \\ \vdots \\ s_j \\ \vdots \\ s_J \end{pmatrix} \tag{5.12}$$

or symbolically

$$\mathbf{B}\boldsymbol{\psi}_{n+1} = \mathbf{C}\boldsymbol{\psi}_n + \mathbf{s}, \tag{5.13}$$

where J is the total length of the spatial mesh (the maximum of j), and the coefficients are

$$B_0 = 1 + 2\frac{D\Delta t}{\Delta x^2}\theta, \qquad B_+ = B_- = -\frac{D\Delta t}{\Delta x^2}\theta, \tag{5.14}$$

$$C_0 = 1 - 2\frac{D\Delta t}{\Delta x^2}(1-\theta), \qquad C_+ = C_- = +\frac{D\Delta t}{\Delta x^2}(1-\theta), \tag{5.15}$$

$$\text{and} \quad s_j = \Delta t[\theta s_j^{(n+1)} + (1-\theta)s_j^{(n)}]. \tag{5.16}$$

[Notice the Δt scaling factor in s_j.]

We are here assuming that the source terms do not depend upon ψ. Eq. (5.13) can be solved formally by inverting the matrix **B**:

$$\boldsymbol{\psi}_{n+1} = \mathbf{B}^{-1}\mathbf{C}\boldsymbol{\psi}_n + \mathbf{B}^{-1}\mathbf{s}. \tag{5.17}$$

Therefore, the additional work involved in using an implicit scheme is that we have (effectively) to invert the matrix **B**, and multiply by the inverse at each timestep.

Provided the mesh is not too large, this can be a manageable approach. In a situation where **B** and **C** do not change with time, the inversion[3] need only be done once; and each step in time involves only a matrix multiplication by $\mathbf{B}^{-1}\mathbf{C}$ of the values from the previous timestep, $\boldsymbol{\psi}_n$.

If the matrix has the tridiagonal form of eq. (5.12), where the entries are non-zero only on the diagonal and the immediately adjacent subdiagonals,

[3] Or, for example, an LU decomposition.

then it will certainly be more computationally efficient to solve for the updated variable by elimination[4] rather than inverting the matrix and multiplying. This is a reflection of the fact that the matrix **B** is *sparse*. All but a very small fraction of its coefficients are zero. The fundamental problem is that the inverse of a sparse matrix is generally not sparse. Consequently, even though multiplication by the original sparse matrix actually requires only a few individual arithmetic operations, and easy short-cuts can be implemented, there are no obvious short-cuts for doing the matrix multiplication by the inverse. As long as we are implementing solutions in a mathematical system like Octave using the built-in facilities, we won't notice any difference because we are not using short-cuts.

5.4 Multiple space dimensions

When there is more than one spatial coordinate dimension, as illustrated in Fig. 5.4, nothing changes formally about our method of solution of parabolic equations. What changes, however, is that we need a systematic way to turn the entire spatial mesh in multiple dimensions into a column vector like those in eq. (5.12). In other words, we must index all the spatial positions with a single index j. But generally if we have multiple dimensions, the natural (physical) indexing of the elements of the mesh is by a number of multiple indices equal to the number of dimensions: e.g. $\psi_{k,l,m}$, where k, l, m correspond to the coordinates x, y, z.

Reordering the mesh elements is not a particularly difficult task algebraically, but it requires an intuitively tricky geometrical mental shift.

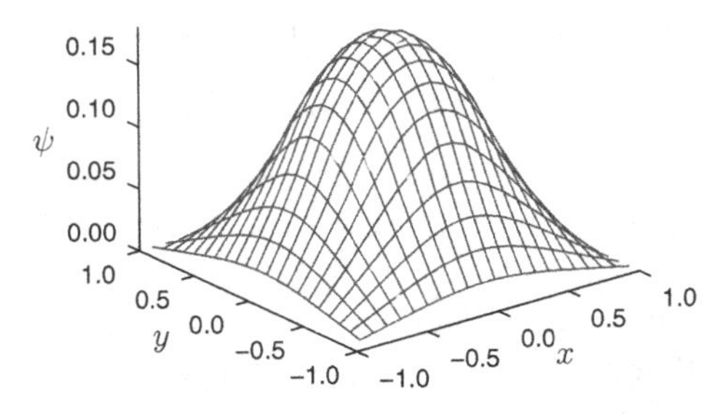

Figure 5.4 Example solution, at a particular time, of a diffusion equation in two space dimensions. The value of ψ is visualized in this perspective plot as the height of the surface.

[4] See for example *Numerical Recipes*, Section 2.6.

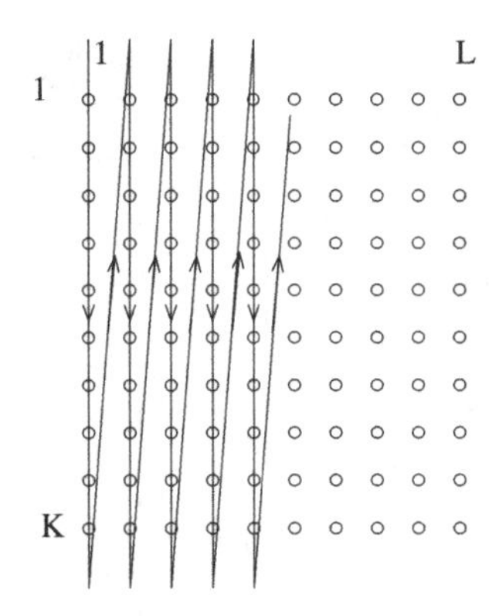

Figure 5.5 Conversion of a two-dimensional array into a single column vector using column order is like stacking all the columns below one another.

In general, as illustated in Fig. 5.5, if we have a quantity $\psi_{k,l,...}$ indexed on a multidimensional mesh whose lengths in the different mesh dimensions are K, L, ... then we re-index them into a single index j as follows. Start with all indices equal to 1. Count through the first index, $k = 1,\ldots,K$. Then increment the second index l and repeat $k = 1,\ldots,K$. Continue this process for $l = 1,\ldots,L$. Then increment the next index (if any), and continue until all indices are exhausted. Incidentally, this is precisely the order in which the elements are stored in computer memory when using a language like Fortran when using the entire allocated array. The result is that for the giant column vector ψ_j, the first K elements are the original multidimensionally indexed elements $\psi_{k,1,...}$, $k = 1,\ldots,K$; the next K elements, $j = K+1,\ldots,2K$, are the original $\psi_{k,2,...}$ and so on.[5]

In multiple dimensions, the second-derivative operator (something like ∇^2) is represented by a stencil, as illustrated in eq. (4.23). The importance of the reordering of the elements into a column vector is that although the components of the stencil that are adjacent in the first index (k) remain adjacent in j, those in the other index(es) (l) do not. For example the elements $\psi_{k,l}$ and $\psi_{k,l+1}$ have j-indices $j = K.(l-1)+k$ and $j = K.l+k$, respectively; they are a distance K apart. This fact means that for multiple spatial dimensions the matrices **B** and **C** are no longer simply tridiagonal. Instead, they have an additional non-zero diagonal, a distance K from the main diagonal. If the boundary conditions are $\psi = 0$, each matrix is "block-tridiagonal," having a form like this:

[5] Octave and Matlab have a simple built-in function to do reordering, called `reshape()`.

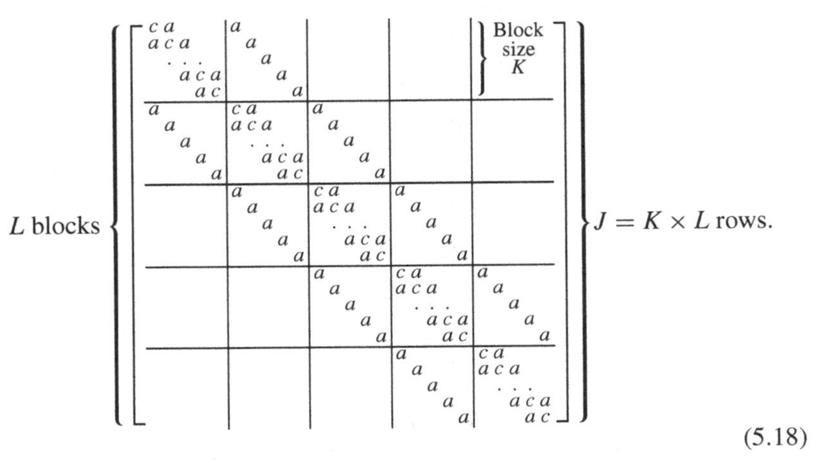

(5.18)

Here, c denotes the coefficient of the *center* of the stencil, and a denotes the coefficients of the *adjacent* points in the stencil. One can think of the total matrix as being composed of $L \times L$ blocks, each of size $K \times K$. All the blocks are zero except the tridiagonal ones. And each non-zero block is itself tridiagonal (or diagonal). If there are further dimensions M, then the giant matrix is a tridiagonal composite of $M \times M$ blocks, each of a two-dimensional ($L \times L$ block) type of eq. (5.18). And so on.

It is very important to think carefully about the boundary conditions. These occur at the boundaries of *each block*. Notice how the corner (boundary) entries of the extra subdiagonal blocks make zero certain coefficients of the subdiagonals of the total matrix. One important consequence of the block-tridiagonal form (5.18) is that it is not so easy to do tridiagonal elimination rather than matrix inversion.

5.5 Estimating computational cost

The computational cost of executing different numerical algorithms often has to be considered for large-scale computing. This cost is generally expressed in the number of floating point operations (FLOPs). Roughly, this is the number of multiplications (or divisions) required. Additions are cheap.

For an $N \times N$ matrix, multiplying into a length-N column vector costs (simplemindedly) N rows times N operations per row: N^2 (FLOPs). By extending this to N columns, we see that multiplying two matrices costs N^3. Although the process of *inversion*[6] of a non-singular matrix seems far more difficult to do, because the algorithm is much more complicated, it also costs

[6] and of LU decomposition and backsubstitution

roughly N^3 operations. These estimates are accurate only to within a factor of two or so. But that is enough for most purposes.

Inverting or multiplying random matrices using Octave on my laptop for $N = 1000$ takes about 1 second. That seems amazingly fast to me, because it corresponds to about 1 FLOP per nanosecond. But that's about what can be achieved these days provided the cache contains enough of the data.

The trouble is that if we are dealing with two dimensions in space, each of length $K = 100$, then the side of the matrix we have to invert is $J = K^2 = 10\,000$. A multiplication or inversion would take at least $J^3/10^9 = 1000$ seconds; that's a quarter of an hour. Waiting that long is frustrating, and if many inversions are needed time rapidly runs out.

Inverting the matrix representing a two-dimensional problem takes K^6 operations. And for a three-dimensional problem it is K^9. This rapid escalation of the cost means that one doesn't generally approach multidimensional problems in terms of simplistic matrix inversion. In particular, it is difficult to implement an implicit scheme for advancing the diffusion equation. And it is probably more realistic just to use an explicit scheme, recognizing and observing the stability limit, eq. (5.6), on maximum timestep size.

Worked example. Crank–Nicholson matrices

Express the following parabolic partial differential equation in two space variables (r and ϕ) and one time variable (t) as a matrix equation using the Crank–Nicholson scheme on uniform grids.

$$\nabla^2\psi = \frac{1}{r^2}\frac{\partial}{\partial r}\left(r^2\frac{\partial\psi}{\partial r}\right) + \frac{1}{r^2}\frac{\partial^2\psi}{\partial\phi^2} = \frac{\partial\psi}{\partial t}. \tag{5.19}$$

Find the coefficients and layout of the required matrices if the boundary conditions are $\partial\psi/\partial r = 0$ at $r = 0$, $\psi = 0$ at $r = R$, and periodicity in ϕ.

Let k and l be indices in the r and ϕ coordinates, and let Δr and $\Delta\phi$ denote their grid spacings. The finite difference form of the spatial derivative terms is

$$\begin{aligned}
\frac{1}{r^2}\frac{\partial}{\partial r}\left(r^2\frac{\partial\psi}{\partial r}\right)_{k,l} &= \frac{1}{r_k^2}\left(r_{k+1/2}^2\frac{\psi_{k+1,l}-\psi_{k,l}}{r_{k+1}-r_k} - r_{k-1/2}^2\frac{\psi_{k,l}-\psi_{k-1,l}}{r_k-r_{k-1}}\right)\frac{1}{r_{k+1/2}-r_{k-1/2}}\\
&= \left[\left(\frac{r_k+\frac{\Delta r}{2}}{r_k}\right)^2(\psi_{k+1,l}-\psi_{k,l}) - \left(\frac{r_k-\frac{\Delta r}{2}}{r_k}\right)^2(\psi_{k,l}-\psi_{k-1,l})\right]\frac{1}{\Delta r^2}\\
&= \left[\left(1+\frac{\Delta r}{2r_k}\right)^2\psi_{k+1,l} - \left(2+\frac{\Delta r^2}{2r_k^2}\right)\psi_{k,l} + \left(1-\frac{\Delta r}{2r_k}\right)^2\psi_{k-1,l}\right]\frac{1}{\Delta r^2}\\
\frac{1}{r^2}\frac{\partial^2\psi}{\partial\theta^2}\bigg|_{k,l} &= \frac{1}{r_k^2}\left(\frac{\psi_{k,l+1}-\psi_{k,l}}{\theta_{l+1}-\theta_l} - \frac{\psi_{k,l}-\psi_{k,l-1}}{\theta_l-\theta_{l-1}}\right)\frac{1}{\theta_{l+1/2}-\theta_{l-1/2}}\\
&= \left(\psi_{k,l+1} - 2\psi_{k,l} + \psi_{k,l-1}\right)\frac{1}{r_k^2\Delta\theta^2}.
\end{aligned} \tag{5.20}$$

Therefore, expressing the positions across the spatial grid in terms of a single index $j = k + K.(l - 1)$ (where K is the size of the k-grid, and L of the l-grid), the differential operator ∇^2 becomes a matrix $\mathbf{M}$ multiplying the vector $\boldsymbol{\psi} = \psi_j$, of the form

$$\mathbf{M} = \left.\left[\begin{array}{c|c|c|c|c}
\begin{smallmatrix} f & e & & & \\ a & c & b & & \\ & \cdot & \cdot & \cdot & \\ & & a & c & b \\ & & & a & c \end{smallmatrix} &
\begin{smallmatrix} d & & & & \\ & d & & & \\ & & d & & \\ & & & d & \\ & & & & d \end{smallmatrix} & & &
\begin{smallmatrix} d & & & & \\ & d & & & \\ & & d & & \\ & & & d & \\ & & & & d \end{smallmatrix} \\ \hline
\begin{smallmatrix} d & & & & \\ & d & & & \\ & & d & & \\ & & & d & \\ & & & & d \end{smallmatrix} &
\begin{smallmatrix} f & e & & & \\ a & c & b & & \\ & \cdot & \cdot & \cdot & \\ & & a & c & b \\ & & & a & c \end{smallmatrix} &
\begin{smallmatrix} d & & & & \\ & d & & & \\ & & d & & \\ & & & d & \\ & & & & d \end{smallmatrix} & & \left.\right\} \begin{smallmatrix} \text{Block} \\ \text{size} \\ K \end{smallmatrix} \\ \hline
 & \begin{smallmatrix} d & & & & \\ & d & & & \\ & & d & & \\ & & & d & \\ & & & & d \end{smallmatrix} &
\begin{smallmatrix} f & e & & & \\ a & c & b & & \\ & \cdot & \cdot & \cdot & \\ & & a & c & b \\ & & & a & c \end{smallmatrix} &
\begin{smallmatrix} d & & & & \\ & d & & & \\ & & d & & \\ & & & d & \\ & & & & d \end{smallmatrix} & \\ \hline
 & & \begin{smallmatrix} d & & & & \\ & d & & & \\ & & d & & \\ & & & d & \\ & & & & d \end{smallmatrix} &
\begin{smallmatrix} f & e & & & \\ a & c & b & & \\ & \cdot & \cdot & \cdot & \\ & & a & c & b \\ & & & a & c \end{smallmatrix} &
\begin{smallmatrix} d & & & & \\ & d & & & \\ & & d & & \\ & & & d & \\ & & & & d \end{smallmatrix} \\ \hline
\begin{smallmatrix} d & & & & \\ & d & & & \\ & & d & & \\ & & & d & \\ & & & & d \end{smallmatrix} & & &
\begin{smallmatrix} d & & & & \\ & d & & & \\ & & d & & \\ & & & d & \\ & & & & d \end{smallmatrix} &
\begin{smallmatrix} f & e & & & \\ a & c & b & & \\ & \cdot & \cdot & \cdot & \\ & & a & c & b \\ & & & a & c \end{smallmatrix}
\end{array}\right]\right\} \text{L blocks.} \tag{5.21}$$

The explanation of this form is as follows. The coefficients of a generic row corresponding to r-position index k are

$$a = \left(1 - \frac{\Delta r}{2r_k}\right)^2 \frac{1}{\Delta r^2}, \quad b = \left(1 + \frac{\Delta r}{2r_k}\right)^2 \frac{1}{\Delta r^2}, \quad d = \frac{1}{r_k^2 \Delta\phi^2}, \quad c = -(a + b + 2d). \tag{5.22}$$

The periodic boundary conditions in ϕ are implemented by the appearance of off-diagonal, d-type, blocks at the upper right and lower left of the matrix. The r-boundary conditions are the same for all the blocks on the diagonal, the tridiagonal acb-type blocks. At the $r = R$ boundary (which would be the bottom row of each block), the condition $\psi = 0$ means no contribution arises to the differential operator from the ψ-value there. The condition therefore allows us simply to omit the $r = R$ row from the matrix, choosing the $k = K$ index to refer to the position $r = R - \Delta r$. At the $r = 0$ end (the top row of each block), the condition $\partial\psi/\partial r = 0$ can be implemented in a properly centered way by choosing the r-grid to be aligned to the half-integral positions.[7] In other words, $r_0 = -\Delta r/2$, $r_1 = +\Delta r/2$, $r_n = (n - 1/2)\Delta r$. (The r_0-position values are not represented in the matrices.) In that case, there is zero contribution to the difference scheme from the first derivative at position $r_{1-1/2} = 0$ (because r^2 there is zero), and the r-second-derivative operator at $k = 1$ becomes

[7] We could alternatively have used integral positions and the second-order scheme of eq. (3.16)

$(1+\frac{\Delta r}{2r_1})^2(\psi_{2,l}-\psi_{1,l})/\Delta r^2 = 4(\psi_{2,l}-\psi_{1,l})/\Delta r^2$, giving rise to equal and opposite coefficients $e = 4/\Delta r^2$. The diagonal entry on those $k = 1$ rows is minus the sum of all the other coefficients on the row: $f = -(e+2d)$.

The Crank–Nicholson scheme for the differential equation time advance is then

$$\boldsymbol{\psi}^{(n+1)} - \boldsymbol{\psi}^{(n)} = \tfrac{\Delta t}{2}\mathbf{M}\boldsymbol{\psi}^{(n+1)} + \tfrac{\Delta t}{2}\mathbf{M}\boldsymbol{\psi}^{(n)}, \tag{5.23}$$

which on rearrangement becomes

$$\boldsymbol{\psi}^{(n+1)} = (\mathbf{I} - \tfrac{\Delta t}{2}\mathbf{M})^{-1}(\mathbf{I} + \tfrac{\Delta t}{2}\mathbf{M})\boldsymbol{\psi}^{(n)}. \tag{5.24}$$

Exercise 5. Diffusion and parabolic equations

1. Write a computer code to solve the diffusive equation

$$\frac{\partial\psi}{\partial t} = D\frac{\partial^2\psi}{\partial x^2} + s(x).$$

for constant, uniform diffusivity D and constant specified source $s(x)$. Use a uniform x-mesh with N_x nodes. Consider boundary conditions to be ψ fixed, equal to ψ_1, ψ_2 at the domain boundaries, $x = -1, 1$, and the initial condition to be $\psi = 0$ at $t = 0$.

Construct a matrix $\mathbf{G} = G_{ij}$ such that $\mathbf{G}\boldsymbol{\psi} = \nabla^2\boldsymbol{\psi}$. Use it to implement the FTCS scheme

$$\boldsymbol{\psi}^{(n+1)} = (\mathbf{I} + \Delta t\mathbf{G})\boldsymbol{\psi}^{(n)} + \Delta t\,\mathbf{s},$$

paying special attention to the boundary conditions.

Solve the time-dependent problem for $D = 1$, $s = 1$, $N_x = 30$, $\psi_1 = \psi_2 = 0$, $t = 0 \to 1$, storing your results in a matrix $\psi(x,t) = \psi_{j_x,j_t}$, and plotting that matrix at the end of the solution, for examination.

Experiment with various Δt to establish the dependence of the accuracy and stability of your solution on Δt. In particular, *without finding an "exact" solution of the equation by any other method*:

(a) Find experimentally the value of Δt above which the scheme becomes unstable.

(b) Estimate experimentally the fractional error arising from finite timestep duration in $\psi(t = 1, x = 0)$ when using a Δt approximately equal to the maximum stable value.

(c) By varying N_x, estimate experimentally the fractional error at $N_x = 30$ arising from finite spatial differences. Which is more significant, time or space difference error?

2. Develop a modified version of your code to implement the θ-implicit scheme

$$(\mathbf{I} - \Delta t \theta \mathbf{G})\psi^{(n+1)} = (\mathbf{I} + \Delta t(1-\theta)\mathbf{G})\psi^{(n)} + \Delta t\, s$$

in the form

$$\psi^{(n+1)} = \mathbf{B}^{-1}\mathbf{C}\psi^{(n)} + \mathbf{B}^{-1}\Delta t\, s.$$

   (a) Experiment with different Δt and θ values for the same time-dependent problem and find experimentally the value of θ for which instability disappears for all Δt.
   (b) Suppose we are limited to only 50 timesteps to solve over the time $0 < t \leq 1$ so $\Delta t = 0.02$. Find experimentally the optimum value of θ which produces the most accurate results.

   Submit the following as your solution for each part:

   1. Your code in a computer format that is capable of being executed, citing the language it is written in.
   2. The requested experimental Δt and/or θ values.
   3. A plot of your solution for at least one of the cases.
   4. A brief description of how you determined the accuracy of the result.

6

Elliptic problems and iterative matrix solution

6.1 Elliptic equations and matrix inversion

In elliptic equations there is no special time-like variable and no preferred direction of propagation of the physical influence. Generally, therefore, elliptic equations, such as Poisson's equation, arise in solving for steady-state conditions in multiple dimensions.

A diffusive problem with constant (time-independent) source (s) and boundary conditions, evolved forward in time, eventually reaches a steady state. When it has reached that state, $\partial/\partial t = 0$. So the steady state satisfies the equation with the time derivative set to zero.

$$\nabla.(D\nabla\psi) = -s. \tag{6.1}$$

This is an elliptic equation.[1] Indeed, if the diffusivity D is uniform, it is just Poisson's equation. The final steady state of a diffusive equation is an elliptic problem.

The linear elliptic problem in space can naturally be framed as a matrix equation by finite differencing in space and expressing the second-order difference operator as a matrix multiplication $\mathbf{B}\boldsymbol{\psi}$ so that

$$\mathbf{B}\boldsymbol{\psi} = -\mathbf{s}. \tag{6.2}$$

This is the matrix inversion problem expressed in standard form. Its solution is (formally)

$$\boldsymbol{\psi} = -\mathbf{B}^{-1}\mathbf{s}. \tag{6.3}$$

So to solve a linear elliptic problem requires simply a matrix inversion. Actually the hard work for the human is expressing the difference equations,

[1] Removing the (first-order) time derivative changes the classification of the equation because it is no longer a differential equation in t at all.

and especially the boundary conditions, in the form of the matrix **B**. But once that's done the computer simply has to invert the matrix. Small problems can readily be solved in this way.

As we've seen for the diffusive problem, though, the matrices involved in multidimensional problems can quickly become overwhelmingly large. The computational cost of inverting them can become excessive. What does one do? Well, we know how to solve a diffusive equation without inverting matrices, don't we? We advance it forward in time using an explicit scheme, being sure to observe the stability limits on timestep. If we take enough timesteps, we'll reach a steady state. Then we'll have the solution to the corresponding elliptic problem.

This is the most appropriate way to solve a gigantic matrix inversion problem. We *do not* invert the matrix. Instead, we *iterate* ψ until it satisfies the matrix equation $\mathbf{B}\boldsymbol{\psi} = -\mathbf{s}$ as accurately as we like, then we have our solution.

How do we iterate? Given what we've said already, we can think of this as the solution of a time-dependent diffusive problem. And for simplicity we'll take the diffusivity uniform:

$$\frac{\partial \psi}{\partial t} = D\nabla^2 \psi + s. \tag{6.4}$$

An iteration takes us from $\psi^{(n)}$ to $\psi^{(n+1)}$. It is essentially a timestep of our diffusive problem. And if we wish to avoid having to invert the matrix, we can use exactly the explicit FTCS scheme to advance it (cf. eq. (5.2)), in one space dimension

$$\psi_j^{(n+1)} - \psi_j^{(n)} = \frac{D\Delta t}{\Delta x^2}(\psi_{j+1}^{(n)} - 2\psi_j^{(n)} + \psi_{j-1}^{(n)} + s_j^{(n)}\Delta x^2/D). \tag{6.5}$$

Observe that the term in parentheses on the right-hand side of this equation is the finite-difference form of the steady-state equation. If ψ satisfies the steady-state equation, then the right-hand side is zero, and there is no change in ψ: $\psi^{(n+1)} = \psi^{(n)}$. If there is no change in ψ, the steady-state equation is satisfied.

Now because we are only really interested in the final steady state, we don't worry about how accurate the time integration is. We just want to get to the final state as fast as we can. However, we do have to worry about stability, because if we use too large a timestep and experience instability because of it, we will probably never reach the steady state. We know the limit of how large Δt can be for this scheme. The limit is $\Delta t \leq \Delta x^2/2D$. If we choose that limit, then the iterative scheme becomes:

$$\psi_j^{(n+1)} - \psi_j^{(n)} = \frac{1}{2}(\psi_{j+1}^{(n)} - 2\psi_j^{(n)} + \psi_{j-1}^{(n)} + \frac{s_j^{(n)}}{D}\Delta x^2). \tag{6.6}$$

In N_d equally-spaced dimensions, where there are $2N_d$ adjacent points in the stencil, the stability limit is $\Delta t = \Delta x^2/2N_dD$ and $2N_d$ replaces 2 in both the leading fraction and the coefficient of $\psi_j^{(n)}$. The general iterative form can be considered to be

$$\psi_j^{(n+1)} - \psi_j^{(n)} = \left(\sum_q a_q \psi_q^{(n)} / \sum_q a_q\right) - \psi_j^{(n)} + \frac{s_j^{(n)} \Delta t}{\sum_q a_q}, \tag{6.7}$$

where the index q ranges over all the $2N_d$ adjacent nodes of the difference stencil. The coefficient of stencil node q is $a_q = D\Delta t/\Delta x^2$. The a_q will not all be the same if the dimensions are unequally spaced. The coefficient of $\psi_j^{(n)}$ is always -1, and exactly cancels the same term on the left-hand side.

This scheme is referred to as Jacobi's method for iterative inversion of matrices.[2] Unfortunately it converges slowly. Fortunately we can do better using schemes inspired by Jacobi's method but using our knowledge more efficiently.

When we are mid way through updating the solution using Jacobi's method, we know some of the new values $\psi_q^{(n+1)}$ at the adjacent nodes. For example, if we were updating in order of increasing indices a two-dimensional spatial configuration, then when we come to update $\psi_{jk}^{(n)}$ we already know $\psi_{j-1,k}^{(n+1)}$ and $\psi_{j,k-1}^{(n+1)}$. Maybe we should use those new values immediately in the difference scheme, rather than the old values at n. Such a scheme of using the currently updated values as available applied to eq. (6.7) is called the Gauss–Seidel method.

6.2 Convergence rate

The Gauss–Seidel method still converges nearly as slowly as the Jacobi method. The easiest way to see this (and the best way to implement the method for partial differential equation solution) is not to update the values in increasing order of index. Instead it is better to update *every other* value. First update all the odd-j values and then update all the even-j values. The advantage is that at each stage of the update all the adjacent values in the stencil that are used have the *same* degree of update. The odd values are updated using the

[2] Jacobi's method actually can be used to solve a rather general matrix equation $\mathbf{B}\boldsymbol{\psi} = \mathbf{s}$, and is implemented as $\boldsymbol{\psi}^{(n+1)} = \mathbf{D}^{-1}(\mathbf{s} - \mathbf{R}\boldsymbol{\psi}^{(n)})$, where $\mathbf{D}$ is the diagonal part of $\mathbf{B}$ and $\mathbf{R} = \mathbf{B} - \mathbf{D}$ is the rest. Its convergence is guaranteed if $\mathbf{B}$ is *diagonally dominant*, meaning that the absolute value of the diagonal term exceeds the sum of the absolute values of all the other coefficients of the corresponding row.

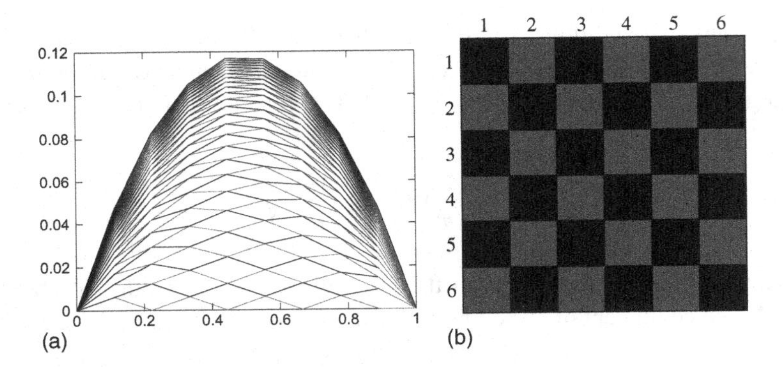

Figure 6.1 (a) One-dimensional Gauss–Seidel odd–even iteration produces successive solutions for each half-step that form a web that progresses upwards toward the solution. (b) In two dimensions the alternate squares to be updated are all the lighter shaded, then all the black.

all old even values. Then the even values are updated using the all new odd values. In multiple dimensions the same effect can be achieved by updating first all the values whose indices sum to an odd value ($j + k + \ldots = \text{odd}$), and then those that sum to an even number. In two dimensions this choice can be illustrated by reference to a checkerboard of red and black squares representing the positions of the nodes. Figure 6.1 illustrates the approach. The algorithm is to update first the red, then the black squares. The red–black updating order separates the update into two half-updates.[3]

Consider a Fourier mode[4] of wave number $k_x = p\pi/L$, where p is the integer mode number and L is the length of the domain (one-dimensional for simplicity) at whose ends Dirichlet boundary conditions are assumed. Its half-update through eq. (6.6) (ignoring the source term), gives rise to an amplification factor

$$A = \psi_j^{(n+1)}/\psi_j^{(n)} = \frac{1}{2}(\mathrm{e}^{ip\pi\,\Delta x/L} + \mathrm{e}^{-ip\pi\,\Delta x/L}) = \cos(p\pi\,\Delta x/L) \tag{6.8}$$

The convergence process consists of the decay of the error in each mode of the system, by which ψ is still different from the steady solution. Each mode is repetitively multiplied by A at each half-step. So after m full-steps ($2m$ half-steps) the mode has decayed by a factor A^{2m}. After some time, the biggest error is going to be caused by the slowest-decaying mode. That is the mode

[3] For the matrix expert, the advancing matrix is then explicitly "two-cyclic, and consistently ordered."

[4] A Fourier treatment is not rigorously justified in general, but serves to illustrate simply the most important characteristics.

whose amplitude factor is closest to 1. Since $A = \cos(p\pi\,\Delta x/L)$, the mode with A closest to 1 is the mode with the smallest wave number, namely $p = 1$. If the number of spatial mesh nodes $N_j = L/\Delta x$ is large, then we can Taylor expand the cosine (for $p = 1$)

$$A \approx 1 - \frac{1}{2}\left(\frac{\pi\,\Delta x}{L}\right)^2 = 1 - \frac{1}{2}\left(\frac{\pi}{N_j}\right)^2. \tag{6.9}$$

To reduce the amplitude of the mode by a factor $1/F$ takes a number of steps m such that $A^{2m} = 1/F$, or, taking the logarithm and using the expansion $\ln(1+x) \approx x$, so $\ln A \approx -\pi^2/2N_j^2$,

$$m = \frac{1}{2}\ln(1/F)/\ln A \approx \ln F\left(\frac{N_j}{\pi}\right)^2. \tag{6.10}$$

This equation shows that to converge by a specified factor requires a number of steps proportional to N_j^2. That is a lot of iterations. Incidentally, the Jacobi iteration obviously leads to the same amplification factor A. The difference is that it is the factor for a full-step, rather than a half-step. Therefore Gauss–Seidel iteration converges only a factor of two faster than Jacobi iteration. On the plus side, for multiple dimensions things don't get significantly worse. A square two-dimensional domain with $\Delta x = \Delta y$ has the same A, so it takes the same number of iterations.

6.3 Successive over-relaxation

The Gauss–Seidel method is a "successive" method where values are updated in succession, and the updated values are immediately used. It turns out that one can greatly improve the convergence rate by the simple expedient of over-correcting the error at each step. This is called "over-relaxation" and when applied to the Gauss–Seidel method is therefore called "successive over-relaxation" or SOR. By analogy with eq. (6.7) it can be written

$$\psi_j^{(n+1)} - \psi_j^{(n)} = \omega\left[\left(\sum_q a_q\psi_q^{(r)} \Big/ \sum_q a_q\right) - \psi_j^{(n)} + \frac{s_j^{(n)}\Delta t}{\sum_q a_q}\right], \tag{6.11}$$

where $r = n$ for q corresponding to odd centered stencils, and $r = n + 1$ for q corresponding to even. The parameter $\omega > 1$ is the over-relaxation parameter. Strictly speaking, if $\omega < 1$ one should speak of under-relaxation. The particular case $\omega = 1$ is the original Gauss–Seidel scheme.

It turns out that SOR is stable for $0 < \omega < 2$. It is intuitively reasonable to guess that SOR converges faster for $\omega > 1$ than for $\omega = 1$. It is not at all straightforward to show how much faster it converges.[5] Therefore we will simply summarize the facts without proving them.

- There is an optimal value of ω somewhere between 1 and 2, where SOR converges fastest.
- If A_J is the amplification factor for the corresponding Jacobi iteration ($\cos(\pi\,\Delta x/L)$ for a uniform problem) then the optimal value is $\omega = \omega_b = 2/(1+\sqrt{1-A_J^2})$.
- For this optimal ω the amplification factor for the SOR is $A_{SOR} = \omega_b - 1 = (A_J\omega_b/2)^2$. For the uniform case and large N_j, these imply

$$\omega_b \approx \frac{2}{1+\pi/N_j}, \qquad \text{and} \quad A_{SOR} \approx 1 - \frac{2\pi}{N_j}. \tag{6.12}$$

These facts show that near the optimal relaxation parameter the number of steps needed to converge by a factor F is approximately $N_j \ln F/2\pi$ (not $N_j^2 \ln F/\pi^2$ as with Gauss–Seidel). That is a very big benefit. However, obtaining that benefit requires one to estimate ω_b accurately, and for more complicated problems doing that becomes hard. Choosing ω well is probably the main challenge for SOR. Fig. 6.2 shows an illustrative example.

There are other iterative matrix solution techniques associated with the name Krylov. Like the SOR solution technique, they use just multiplication by the matrix, not inversion. That is a big advantage for very sparse matrices arising in partial differential equation solving. They go by names like "Conjugate Gradient", "Newton Krylov," and "GMRES". In some situations they converge faster than SOR, and they don't require careful adjustment of a relaxation parameter. However, they have their own tuning problems associated with "preconditioning." These topics are introduced very briefly in the final chapter.

6.4 Iteration and non-linear equations

A major advantage accrues to iterative methods of solving elliptic problems. It is that we only have to *multiply by* the difference matrix. Because that difference matrix is extremely sparse, we need never in fact construct it in its entirety. We perform the matrix multiplication more physically by

[5] See the enrichment section for an outline if you are interested. G. D. Smith (1985) *Numerical Solution of Partial Differential Equations*, Oxford University Press, Oxford, p. 275ff gives an accessible detailed treatment.

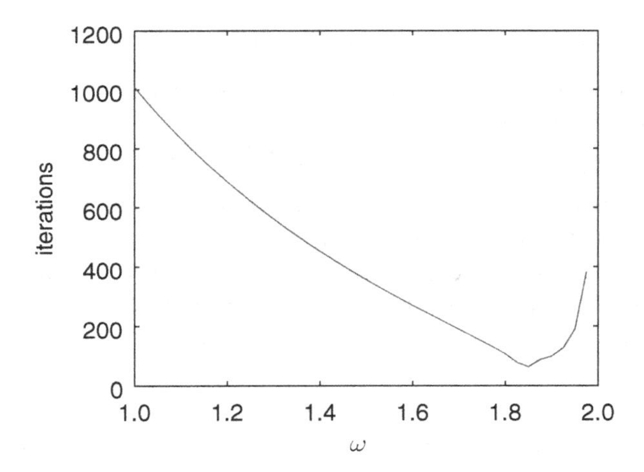

Figure 6.2 Number of iterations required to converge an SOR solution of Poisson's equation with uniform source on a mesh of length $N_j = 32$. It is declared converged when the maximum ψ change in a step is less than $10^{-6}\psi_{\max}$. The minimum number of iterations is found to be 63 at $\omega = 1.85$. This should be compared with theoretical values of $\ln(10^6)(N_j/2\pi) = 70$ at $\omega = 2/(1+\pi/N_j) = 1.821$.

performing the small number of multiplications of adjacent stencil mesh values by coefficients. This saves immense amounts of storage space (compared with constructing the full matrix) and immense numbers of irrelevant multiplications by zero. The cost we must pay for the benefit of not constructing and inverting the matrix is that a substantial number of iterations is necessary. Generally, for sizable multidimensional problems, that iteration cost is far less than the savings.

Another situation where iteration becomes essential is when the differential equations are *non-linear*. In the screening of electric fields by plasmas or electrolytes, for example, the source of the Poisson equation (the charge density) is a non-linear function of the potential ϕ, in this case an exponential leading to

$$\nabla^2\phi = -s = \exp(\phi) - 1. \tag{6.13}$$

How do we solve such an elliptic equation? It cannot be solved by matrix inversion because of the non-linearity on the right-hand side. Even if we invert the difference matrix and construct eq. (6.3): $\boldsymbol{\phi} = -\mathbf{B}^{-1}\mathbf{s}$, because s is a non-linear function of ϕ, this expression does not constitute a solution of the problem.

The generic answer to how to solve a non-linear problem is:

1. Linearize it about the current estimate of the solution.
2. Solve or advance the linear problem.
3. Repeat from 1, until converged on the non-linear solution.

When one knows that iteration is required anyway, because the problem is non-linear, one has a substantial incentive to use an iterative method for the linearized part of the problem as well. Very often it will cost no more effort to solve a non-linear problem by iteration than a linear problem.

6.4.1 Linearization

Suppose we have the potential function $\phi^{(n)}$, at step n, not yet a solution of the steady elliptic equation. In the neighborhood of that function, we can express the source via a Taylor expansion (at each point in the mesh):

$$s(\phi) = s(\phi^{(n)}) + \frac{\partial s}{\partial \phi}\tilde{\phi} + \frac{1}{2!}\frac{\partial^2 s}{\partial \phi^2}\tilde{\phi}^2 \ldots, \qquad \text{where} \quad \tilde{\phi} = (\phi - \phi^{(n)}). \tag{6.14}$$

For values of ϕ close enough to $\phi^{(n)}$, i.e. small enough $\tilde{\phi}$, we can ignore all but the first two terms of this expansion. For our exponential example, substituting for ϕ and rearranging the terms, we would then obtain the linearized equation

$$\nabla^2\tilde{\phi} - \exp(\phi^{(n)})\tilde{\phi} = \exp(\phi^{(n)}) - 1 - \nabla^2\phi^{(n)}. \tag{6.15}$$

The right-hand side is the residual – the amount by which the step-n value fails to satisfy the differential equation. More generally one obtains, from $\nabla^2\phi = -s$, a linearized equation

$$\nabla^2\tilde{\phi} + \left.\frac{\partial s}{\partial \phi}\right|_{\phi^{(n)}} \tilde{\phi} = -s(\phi^{(n)}) - \nabla^2\phi^{(n)}. \tag{6.16}$$

This equation can be solved to find $\tilde{\phi}$ linearly. Of course $\tilde{\phi}$ is our linearized estimate of the amount by which the real solution ϕ is different from the starting value at the nth step: $\phi^{(n)}$. If we find the solution of the *linearized* equation (6.16), then, since that equation is only approximate, the new value $\phi^{(n+1)} \equiv \phi^{(n)} + \tilde{\phi}$ is only an approximate solution of the original non-linear equation. Presumably, though, it is closer to the actual solution. So if we simply interate the process, as n increases we approach the full non-linear solution. That is an iterative solution to the non-linear problem using linearization.

6.4.2 Combining linear and non-linear iteration

The question then arises: what method should we use to solve the linear problem to find $\tilde{\phi}$? It is natural to be guided by the knowledge that even if we solved the linear problem exactly, that would not give us an exact solution of the nonlinear problem. Therefore it is *unnecessary* to solve the linearized equation exactly. And in many cases it is in fact not even necessary to get close to the exact solution of the linearized equation for each interation of the non-linear equation. What we then can do is to say, we'll solve the linearized equation iteratively, but we'll use *only one step* in our solution. In other words, $\phi^{(n+1)}$ is arrived at by doing a single advance of the linearized equation iterative scheme (e.g. an SOR advance). Then we recalculate s and its derivative $\partial s/\partial\phi$ for the new value of ϕ, and iterate.

Actually, one might sometimes be able to dispense with the linear term in the linearization, by retaining, in the Taylor expansion for s, only the first, constant term. That would amount to using $\nabla^2\phi^{(n+1)} = \exp\phi^{(n)} - 1$ as the equation to be solved for each step of the non-linear iteration. Whether that will work depends upon the relative importance of the $\nabla^2\phi$ term in the equation. In places where $\nabla^2\phi$ is small, the non-linear equation behaves like a transcendental equation for ϕ: simply $\exp\phi - 1 \approx 0$. In that case, solving eq. (6.15) (without the $\nabla^2\phi$ term) is equivalent to a single Newton iteration of a root-finding problem, which is a sensible iteration to take. Without the linear term, though, negligibly small advance towards the non-linear solution will occur.

It is hard to generalize about how fast the iteration is going to converge on the solution of the non-linear equation. It depends upon the type of non-linearity. But it is usually the case that if the iterative advance is chosen reasonably, then the convergence to the non-linear solution takes no more iterations than approximately what it would require to converge as accurately to a solution of the linearized equations. In short, iterative solutions can readily accommodate non-linearity in the equations, and produce solutions with comparable computational cost.

Worked example. Optimal SOR relaxation

Consider the elliptic equation

$$\frac{\partial^2\phi}{\partial x^2} + \frac{\partial^2\phi}{\partial y^2} = s(x, y) \tag{6.17}$$

expressed on a cartesian grid $x = j\Delta x$, $j = 0, 1, \ldots, N_x$; $y = k\Delta y$, $k = 0, 1, \ldots, N_y$. And suppose the boundary conditions at $x = 0$ and $N_x\Delta x$, and

$y = 0$ and $N_y\Delta y$, are $\phi = f(x, y)$. Find the optimum relaxation parameter ω for an SOR iterative solution of the system, and the resulting convergence rate.

Suppose the final solution of the system is denoted ϕ_s. We can define a new dependent variable $\psi = \phi - \phi_s$, which is the error between some approximation of the solution (ϕ) and the actual solution. Of course, while we are in the process of finding the solution, we don't know how to derive ψ from ϕ, because we don't yet know what ϕ_s is. That fact does not affect the following arguments. Substituting for $\phi = \psi + \phi_s$ in the differential equation and using the fact that ϕ_s exactly satisfies it and the boundary conditions, we immediately deduce that ψ satisfies the homogeneous differential equation

$$\frac{\partial^2\psi}{\partial x^2} + \frac{\partial^2\psi}{\partial y^2} = 0, \tag{6.18}$$

together with homogeneous boundary conditions: $\psi = 0$ on the boundary. Of course the final solution for ψ is, as a consequence, simply zero. But prior to arriving at that solution, ψ is non-zero and any iteration scheme that we use to solve for ϕ is equivalent to an iteration scheme to solve for ψ. In particular, the convergence rate of ψ to zero is just the convergence rate of ϕ to ϕ_s. All stability and convergence analysis can be done on the simpler homogeneous ψ system (6.18), and its results applied immediately to the ϕ system (6.17).

The homogeneous Jacobi iteration (see eq. (6.6)) in two dimensions of different grid spacing is

$$\begin{aligned}\psi_{j,k}^{(n+1)} &- \psi_{j,k}^{(n)} \\ &= \frac{1}{2}\left(\frac{\psi_{j+1,k}^{(n)} + \psi_{j-1,k}^{(n)}}{\Delta x^2} + \frac{\psi_{j,k+1}^{(n)} + \psi_{j,k-1}^{(n)}}{\Delta y^2}\right)\left(\frac{1}{\Delta x^2} + \frac{1}{\Delta y^2}\right)^{-1} - \psi_{j,k}^{(n)}.\end{aligned} \tag{6.19}$$

Now we do a Von Neumann analysis of the homogeneous system, examining the Fourier modes of the two-dimensional system. They are proportional to $\exp i(k_x x + k_y y) = \exp i\pi(pj/N_x + qk/N_y)$, where p and q are integers that label the mode.[6] For the p, q Fourier mode, $\psi_{j+1,k}^{(n)} + \psi_{j-1,k}^{(n)} = 2\cos(p\pi/N_x)\psi_{j,k}^{(n)}$ and $\psi_{j,k+1}^{(n)} + \psi_{j,k-1}^{(n)} = 2\cos(q\pi/N_y)\psi_{j,k}^{(n)}$. So, substituting, we get the two-dimensional version of eq. (6.8)

$$A_J \equiv \psi_{j,k}^{(n+1)}/\psi_{j,k}^{(n)} = \left(\frac{\cos(p\pi/N_x)}{\Delta x^2} + \frac{\cos(q\pi/N_y)}{\Delta y^2}\right)\left(\frac{1}{\Delta x^2} + \frac{1}{\Delta y^2}\right)^{-1}. \tag{6.20}$$

[6] If we apply the boundary conditions, then the *eigenmode* is actually $\sin(jp\pi/N_x)\sin(kq\pi/N_y)$, but the complex Von Neumann analysis gives the same result.

The slowest-decaying mode is the longest wavelength mode: $p = 1, q = 1$. For this mode, expanding $\cos\theta \approx 1 - \theta^2/2$, we get

$$\begin{aligned} A_J &\approx 1 - \left(\frac{\pi^2}{2N_x^2\Delta x^2} + \frac{\pi^2}{2N_y^2\Delta y^2}\right)\left(\frac{1}{\Delta x^2} + \frac{1}{\Delta y^2}\right)^{-1} \\ &= 1 - \frac{1}{2}\left[\left(\frac{\pi}{N_x}\right)^2 \frac{\Delta y^2}{\Delta x^2 + \Delta y^2} + \left(\frac{\pi}{N_y}\right)^2 \frac{\Delta x^2}{\Delta x^2 + \Delta y^2}\right] \\ &= 1 - \frac{1}{2}\left(\frac{\pi}{N_x N_y}\right)^2 \frac{N_x^2\Delta y^2 + N_y^2\Delta x^2}{\Delta x^2 + \Delta y^2}. \end{aligned} \tag{6.21}$$

For brevity in the rest of our equations, let's define a number to represent the second term

$$M \equiv N_x N_y \Big/ \sqrt{\frac{N_x^2\Delta y^2 + N_y^2\Delta x^2}{\Delta x^2 + \Delta y^2}} \tag{6.22}$$

so that $A_J = 1 - \frac{1}{2}\left(\frac{\pi}{M}\right)^2$ and M serves as a measure of the grid size, like N in the one-dimensional problem.[7] The optimal SOR relaxation factor ω_b is then expressed in terms of A_J as

$$\omega_b = \frac{2}{1 + \sqrt{(1 + A_J)(1 - A_J)}} \approx \frac{2}{1 + \frac{\pi}{M}}. \tag{6.23}$$

The resulting amplification factor for SOR iteration using this ω_b is

$$A_{SOR} = \omega_b - 1 \approx 1 - \frac{2\pi}{M}, \tag{6.24}$$

and the number of iterations required to reduce ψ by a factor $1/F$ is (cf. 6.10)

$$m = -\ln F / \ln A_{SOR} \approx M \frac{\ln F}{2\pi}. \tag{6.25}$$

Enrichment: Outline of SOR convergence analysis

Assume the matrix $\mathbf{B}$ in eq. (6.2) to be solved is arbitrary except that its diagonal entries are minus unity. That can be ensured without loss of generality by scaling the equations. It can then be separated into three parts: the diagonal, which is just minus the unit matrix $\mathbf{I}$, plus $\mathbf{U}$, those entries that multiply the old ψ-values (even nodes), plus $\mathbf{L}$, those entries that multiply the new values (odd nodes). $\mathbf{B} = -\mathbf{I} + \mathbf{U} + \mathbf{L}$. Then the SOR scheme (ignoring source) can be written

$$\boldsymbol{\psi}^{(n+1)} - \boldsymbol{\psi}^{(n)} = \omega[(-\mathbf{I} + \mathbf{U})\boldsymbol{\psi}^{(n)} + \mathbf{L}\boldsymbol{\psi}^{(n+1)}].$$

Collecting n terms together,

$$(\mathbf{I} - \omega\mathbf{L})\boldsymbol{\psi}^{(n+1)} = [(1 - \omega)\mathbf{I} + \omega\mathbf{U}]\boldsymbol{\psi}^{(n)},$$

[7] In fact, if $N_x = N_y$ and $\Delta x = \Delta y$, then $M = N_x$, or alternatively if $\Delta y \gg \Delta x$, and N_x is not far smaller than N_y, then $M \approx N_x$.

which can be written

$$\psi^{(n+1)} = (\mathbf{I} - \omega\mathbf{L})^{-1}[(1-\omega)\mathbf{I} + \omega\mathbf{U}]\psi^{(n)} = \mathbf{H}\psi^{(n)}.$$

The eigenvalues of the advancing matrix $\mathbf{H}$ are the "amplification factors" for the true modes of the system. They are the solutions, λ, of $\det(\mathbf{H} - \lambda\mathbf{I}) = 0$. But

$$\mathbf{H} - \lambda\mathbf{I} = (\mathbf{I} - \omega\mathbf{L})^{-1}\{(1-\omega)\mathbf{I} + \omega\mathbf{U} - \lambda(\mathbf{I} - \omega\mathbf{L})\}.$$

So

$$\det\{\lambda\omega\mathbf{L} + (1 - \lambda - \omega)\mathbf{I} + \omega\mathbf{U}\} = 0.$$

Now the determinant of any matrix $\alpha^{-1}\mathbf{L} - \mathbf{D} + \alpha\mathbf{U}$, where $\mathbf{L}$ and $\mathbf{U}$ are lower and upper triangular parts and $\mathbf{D}$ is the diagonal, is independent of α. This can be seen by noticing that any term in the expansion of the determinant has equal numbers of elements from $\mathbf{U}$ as it has from $\mathbf{L}$; so the α factors cancel out. As implied by our notation, we can arrange the nodes in an appropriate order such that all the even coefficients are in the upper triangle and the odd coefficients in the lower triangle part of the matrix. This would be achieved by the simple expedient of putting all the even positions first. Actually we don't need to do the rearrangement. We just need to know it could be done. In that case, we can balance the upper and lower parts of the determinantal equation, multiplying the $\mathbf{L}$ term by $\lambda^{-1/2}$ and the $\mathbf{U}$ term by $\lambda^{1/2}$ to make it:

$$\det\{\lambda^{1/2}\omega\mathbf{L} + (1 - \lambda - \omega)\mathbf{I} + \lambda^{1/2}\omega\mathbf{U}\} = 0,$$

i.e.

$$\det\{-(\lambda + \omega - 1)\omega^{-1}\lambda^{-1/2}\mathbf{I} + \mathbf{L} + \mathbf{U}\} = 0.$$

Now notice that the eigenvalues μ for the Jacobi iteration matrix, $\mathbf{L}+\mathbf{U}$, satisfy $\det(-\mu\mathbf{I}+\mathbf{L}+\mathbf{U}) = 0$, which is exactly the same equation with the identification $(\lambda + \omega - 1)\omega^{-1}\lambda^{-1/2} = \mu$. There's a direct mapping between eigenvalues of the Jacobi iteration and of the SOR iteration.

The relationship can be considered a quadratic equation for λ, given μ and ω

$$\lambda^2 + (2\omega - 2 - \omega^2\mu^2)\lambda + (\omega - 1)^2 = 0.$$

The optimum ω gives the smallest magnitude of the larger λ solution. It occurs when the λ roots coincide, i.e. when $(\omega - 1 - \omega^2\mu^2/2)^2 = (\omega - 1)^2$ whose solution is

$$\omega = \omega_b = \frac{2}{1 + \sqrt{1 - \mu^2}}.$$

The corresponding eigenvalue is $\lambda = \omega_b - 1$. For $\omega > \omega_b$, the roots for λ are complex with magnitude $\omega - 1$. Therefore SOR is stable only for $\omega < 2$, and the convergence rate degrades linearly to zero between ω_b and 2.

Exercise 6. Iterative solution of matrix problems

1. Start with your code that solved the diffusion equation explicitly. Adjust it to always take timesteps at the stability limit $\Delta t = \Delta x^2/2D$, so that

$$\psi_j^{(n+1)} - \psi_j^{(n)} = \left(\frac{1}{2}\psi_{j+1}^{(n)} - \psi_j^{(n)} + \frac{1}{2}\psi_{j-1}^{(n)} + \frac{s_j^{(n)}}{2D}\Delta x^2\right).$$

Now it is a Jacobi iterator for solving the steady-state elliptic matrix equation. Implement a convergence test that finds the maximum absolute *change in* ψ and divides it by the maximum absolute ψ, giving the normalized ψ-change. Consider the iteration to be converged when the normalized ψ-change is less than (say) 10^{-5}. Use it to solve

$$\frac{d^2\psi}{dx^2} = 1$$

on the domain $x = [-1, 1]$ with boundary conditions $\psi = 0$, with a total of N_x equally-spaced nodes. Find how many iterations it takes to converge, starting from an initial state $\psi = 0$, when

(a) $N_x = 10$,
(b) $N_x = 30$,
(c) $N_x = 100$.

Compare the number of iterations you require with the analytic estimate in the notes. How good is the estimate?

Now we want to check how accurate the solution really is.

(d) Solve the equation analytically, and find the value of ψ at $x = 0$, $\psi(0)$.
(e) For the three N_x values, find the relative error[8] in $\psi(0)$.
(f) Is the actual relative error the same as the convergence test value 10^{-5}? Why?

2. Optional and not for credit. Turn your iterator into an SOR solver by splitting the iteration matrices up into dark gray and black (odd and even) advancing parts. Each part-iterator then uses the latest values of ψ, that has just been updated by the other part-iterator. Also, provide yourself an over-relaxation parameter ω. Explore how fast the iterations converge as a function of N_x and ω.

Note. Although in Octave/Matlab it is convenient to implement the matrix multiplications of the advance using a literal multiplication by a big sparse matrix, one does not do that in practice. There are far more efficient ways of doing the multiplication that avoid all the irrelevant multiplications by zero.

[8] The difference between the "converged" iterative $\psi(0)$ and the analytic $\psi(0)$ normalized to the analytic value.

7

Fluid dynamics and hyperbolic equations

7.1 The fluid momentum equation

In Section 4.1.1 we introduced the continuity or mass-conservation equation of fluid flow:

$$\frac{\partial \rho}{\partial t} + \nabla.(\rho \boldsymbol{v}) = S. \tag{7.1}$$

Now we want to discuss the second important equation governing fluid dynamics, the momentum-conservation equation. Like mass conservation, momentum conservation simply identifies all the sources of momentum within a particular volume V and the fluxes of momentum inward across the boundary ∂V of that volume, and sets their sum equal to the rate of change of the total momentum in the volume. Momentum is of course a vector quantity whose density (momentum per unit volume) is $\rho \boldsymbol{v}$. The total rate of change of momentum is the integral of this quantity over the volume.

The sources of momentum within a volume consist of any *body forces* that might be acting upon the fluid. This, of course, is what Newton's second law of motion tells us. Rate of change of momentum is equal to force. However, like the momentum, the force must be expressed in terms of *force density* $\boldsymbol{F}$, the force per unit volume acting on the fluid. For example, gravity gives rise to a force per unit volume $\rho \boldsymbol{g}$, where $\boldsymbol{g}$ is the gravitational acceleration vector (downwards on Earth). Or again, if the fluid is electrically charged with a charge density ρ_q, then the body force density arising from an electric field $\boldsymbol{E}$ is $\rho_q \boldsymbol{E}$. The force density $\boldsymbol{F}$ is the sum of all such forces that happen to be present. There might be none.

The flux of momentum across the surface is the more tricky part. Some of that flux arises because of fluid motion. The fluid momentum, density $\rho \boldsymbol{v}$, is being carried along, "convected," with the fluid at velocity $\boldsymbol{v}$. Consequently, across any stationary surface element $\boldsymbol{dA}$ there is a convective flux of momentum equal to $\rho \boldsymbol{vv}.\boldsymbol{dA}$. We may therefore identify the convective momentum flux density as the quantity whose dot product with $\boldsymbol{dA}$ gives the flux across $\boldsymbol{dA}$. It is $\rho \boldsymbol{vv}$, which is a tensor (or dyadic in this notation), it has two sets of coordinate indices, and can be thought of as a 3×3 matrix:

$$\rho \boldsymbol{vv} = \rho v_i v_j = \rho \begin{pmatrix} v_1 v_1 & v_1 v_2 & v_1 v_3 \\ v_2 v_1 & v_2 v_2 & v_2 v_3 \\ v_3 v_1 & v_3 v_2 & v_3 v_3 \end{pmatrix}. \tag{7.2}$$

In addition to this convective momentum flux, carried by the local mean fluid velocity, there may be momentum flux that arises from other effects. One such effect is pressure. Another is viscosity. Another (in non-Newtonian fluids like gels or of course solids) might be shear stress arising from elasticity. All of these can be lumped together into another tensor that is usually called simply the *stress* tensor, or the *pressure* tensor. We'll write it $\mathbf{P}$. It is a 3×3 matrix with coefficients P_{ij}. We assume that just as $\boldsymbol{F}$ is the sum of all possible body force densities, $\mathbf{P}$ is the sum of all non-convective momentum flux densities.

The conservation of momentum is then

$$\frac{\partial}{\partial t}\int_V \rho \boldsymbol{v} d^3x = \int_V \boldsymbol{F} d^3x - \int_{\partial V} (\rho \boldsymbol{vv} + \mathbf{P}).\boldsymbol{dA}, \tag{7.3}$$

applied to an arbitrary volume and surface, as illustrated in Fig. 7.1. Just as with mass conservation, we can use Gauss's divergence theorem to turn the surface integral into a volume integral, and gather the terms together:

$$\int_V \frac{\partial}{\partial t}(\rho \boldsymbol{v}) - \boldsymbol{F} + \nabla.(\rho \boldsymbol{vv} + \mathbf{P}) d^3x = 0. \tag{7.4}$$

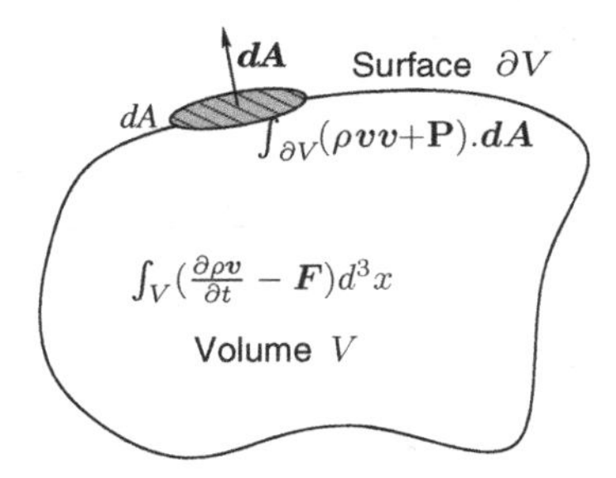

Figure 7.1 Integral of momentum flux density across the boundary surface ∂V is equal to minus the integral of rate of change of momentum minus force density over the volume V. The momentum flux density includes convective flux and stress tensor parts.

This equation must hold for any volume V, and the only way for that to be true is for the integrand to be zero everywhere:

$$\frac{\partial}{\partial t}(\rho \boldsymbol{v}) - \boldsymbol{F} + \nabla.(\rho \boldsymbol{v}\boldsymbol{v} + \mathbf{P}) = 0. \tag{7.5}$$

This is the general form of the fluid momentum conservation equation. If we know what **P** is, then this equation is enough to solve for $\boldsymbol{v}$. But really we are in the same situation as we were with the continuity equation. There, we could solve the equation for ρ, but only if we knew $\boldsymbol{v}$. Now we've got an equation for $\boldsymbol{v}$, but it depends upon knowing **P**. Intuitively you can see that this heirarchical process might go on for ever. We can get an equation for **P** from the conservation of *energy*, but that equation will contain a third-order tensor governing the energy flux (conduction etc.). Solving for that requires yet another equation, and so on. In general, to get a soluble problem we have to call a halt at some point – a process called "closure." How and when we do that decides what sort of fluid equations we end up with. This closure generally invokes a "constitutive relation" between a property such as stress and the other variables of the fluid such as density or velocity gradient.

The kinds of fluids we encounter most in everyday life are isotropic. They have no intrinsically preferred direction. There are fluids that are anisotropic, for example plasmas or other electrically conducting fluids in magnetic fields. But for now we set them aside. Isotropic fluids generally give rise to nearly symmetric stress tensors **P**. It is then helpful to separate out the total stress tensor into a part that is simply a scalar p times the unit matrix **I** (that's the isotropic part), and a part $\boldsymbol{\sigma}$ that has zero *trace*, i.e. the sum of its diagonal elements is zero $\sum_i \sigma_{ii} = 0$. Thus we write $\mathbf{P} = p\mathbf{I} + \boldsymbol{\sigma}$. Then p is the *pressure*. The traceless stress tensor $\boldsymbol{\sigma}$ for simple fluids arises from viscosity, which relates stress to the rate of strain tensor, see Fig. 7.2. The rate of strain tensor is

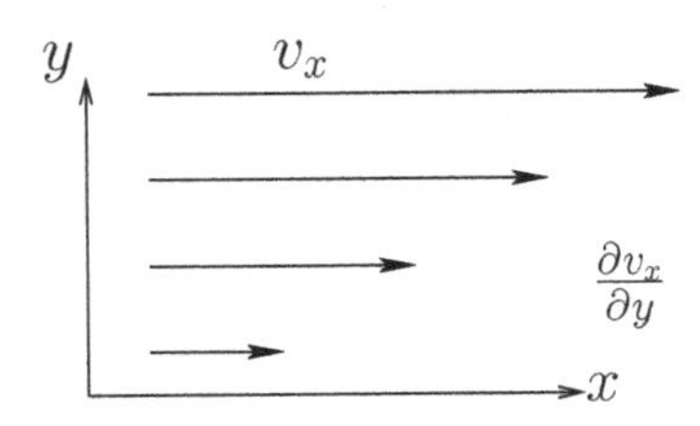

Figure 7.2 The transfer in the y-direction of x-momentum arises from the rate of strain dv_x/dy. The rate of strain tensor is the symmetric generalization of this form.

$$\frac{1}{2}\left(\frac{\partial v_i}{\partial x_j}+\frac{\partial v_j}{\partial x_i}\right).$$

And $\boldsymbol{\sigma}$ is proportional to its traceless part

$$\sigma_{ij}=\mu\left[\left(\frac{\partial v_i}{\partial x_j}+\frac{\partial v_j}{\partial x_i}\right)-\frac{2}{3}\nabla.\boldsymbol{v}\delta_{ij}\right]=\mu\left[\left((\nabla\boldsymbol{v})+(\nabla\boldsymbol{v})^T\right)-\frac{2}{3}(\nabla.\boldsymbol{v})\mathbf{I}\right]_{ij}. \tag{7.6}$$

The constant of proportionality μ is the (shear) *viscosity*. Here $(\nabla\boldsymbol{v})$ is a tensor (dyadic) whose transpose is indicated with a superscript T. Substituting this expression into the general momentum-conservation equation gives what is called the Navier–Stokes equation:

$$\frac{\partial}{\partial t}(\rho\boldsymbol{v})+\nabla.(\rho\boldsymbol{v}\boldsymbol{v})=-\nabla.(p\mathbf{I}+\boldsymbol{\sigma})+\boldsymbol{F}=-\nabla p-\mu\nabla^2\boldsymbol{v}-\frac{1}{3}\mu\nabla(\nabla.\boldsymbol{v})+\boldsymbol{F}. \tag{7.7}$$

The closure for the pressure (and viscosity) must generally be determined by equations of state relating pressure p to density ρ; for example for an ideal isothermal gas $p\propto\rho$. Liquids have an equation of state that amounts approximately to incompressibility, $\rho = const.$, and they generally have a zero volumetric source S. For such a fluid, the continuity equation shows that the velocity divergence is zero, $\nabla.\boldsymbol{v}=0$. The divergenceless fluid momentum equation is then simpler.

$$\frac{\partial}{\partial t}(\rho\boldsymbol{v})+\nabla.(\rho\boldsymbol{v}\boldsymbol{v})=-\nabla p-\mu\nabla^2\boldsymbol{v}+\boldsymbol{F}. \tag{7.8}$$

And of course if viscosity and body forces are negligible it becomes even simpler yet.

The left-hand side of these equations is often rewritten using the continuity equation with $S=0$ to demonstrate

$$\frac{\partial}{\partial t}(\rho\boldsymbol{v})+\nabla.(\rho\boldsymbol{v}\boldsymbol{v})=\rho\left(\frac{\partial}{\partial t}\boldsymbol{v}+\boldsymbol{v}.\nabla\boldsymbol{v}\right). \tag{7.9}$$

Then the second form is recognized as ρ times the convective derivative $\partial/\partial t+\boldsymbol{v}.\nabla$ of $\boldsymbol{v}$. However, the first form is what is called "conservative" form, and it is by far the better form to use for discrete representation and numerical solution on fixed meshes.

7.2 Hyperbolic equations

Fluid equations are generally hyperbolic. Let's start our analysis of such hyperbolic equations by considering a problem where body force is zero; sources are zero; viscosity is zero; pressure is related to density by an adiabatic law $p\rho^{-\gamma} = const.$; and the configuration is one-dimensional in space. This is governed then by the following equations:

$$
\begin{aligned}
&\text{continuity:} && \frac{\partial \rho}{\partial t} + \frac{\partial}{\partial x}(\rho v) = 0; \\
&\text{momentum:} && \frac{\partial}{\partial t}(\rho v) + \frac{\partial}{\partial x}(\rho v^2) = -\frac{\partial}{\partial x}p\,; \\
&\text{state:} && p\rho^{-\gamma} = const.
\end{aligned}
\tag{7.10}
$$

These are three equations for three unknowns ρ, v, and p. They represent a compressible fluid (gas) in a pipe, for example. We can eliminate p immediately by writing $p = p_0\rho^\gamma/\rho_0^\gamma$. To retain the conservation properties, we regard the density ρ and *momentum density*, $\rho v = \Gamma$, as the independent variables, in which case the equations become

$$
\begin{aligned}
\frac{\partial \rho}{\partial t} &= -\frac{\partial \Gamma}{\partial x} \\
\frac{\partial \Gamma}{\partial t} &= -\frac{\partial}{\partial x}(\Gamma^2/\rho + (p_0/\rho_0^\gamma)\rho^\gamma).
\end{aligned}
\tag{7.11}
$$

We might well want to solve these non-linear equations numerically. They are now expressed in a form that is actually the same for all types of fluid conservation equations:

$$
\frac{\partial \mathbf{u}}{\partial t} = -\frac{\partial \mathbf{f}}{\partial x}. \tag{7.12}
$$

In our particular case

$$
\mathbf{u} = \begin{pmatrix} \rho \\ \Gamma \end{pmatrix} \quad \text{and} \quad \mathbf{f} = \begin{pmatrix} \Gamma \\ \Gamma^2/\rho + (p_0/\rho_0^\gamma)\rho^\gamma \end{pmatrix} \tag{7.13}
$$

are the state vector and the flux vector, respectively. Since the flux vector is a function of the state vector, we can use the chain rule to write the equations as

$$
\frac{\partial \mathbf{u}}{\partial t} = -\frac{\partial \mathbf{f}}{\partial \mathbf{u}}\frac{\partial \mathbf{u}}{\partial x} = -\sum_{m=1,M}\frac{\partial \mathbf{f}}{\partial u_m}\frac{\partial u_m}{\partial x} = -\mathbf{J}\frac{\partial \mathbf{u}}{\partial x}. \tag{7.14}
$$

Here, the Jacobian matrix $\mathbf{J} = \partial\mathbf{f}/\partial\mathbf{u}$ has size $M \times M = 2 \times 2$ and is explicitly

$$\mathbf{J} = \begin{pmatrix} 0 & 1 \\ -\Gamma^2/\rho^2 + \gamma(p_0/\rho_0^\gamma)\rho^{\gamma-1} & 2\Gamma/\rho \end{pmatrix}. \tag{7.15}$$

The Jacobian matrix quite generally embodies the differential equation by relating time-derivates to space-derivatives of the state vector, through eq. (7.14):

$$\frac{\partial \mathbf{u}}{\partial t} = -\mathbf{J}\frac{\partial \mathbf{u}}{\partial x}.$$

Writing for convenience $\Gamma/\rho = v$, and $\gamma(p_0/\rho_0^\gamma)\rho^{\gamma-1} = c_s^2$, the eigenvalues of $\mathbf{J}$ are then solutions of

$$\begin{vmatrix} -\lambda & 1 \\ -v^2 + c_s^2 & -\lambda + 2v \end{vmatrix} = \lambda^2 - 2v\lambda + v^2 - c_s^2 = 0, \tag{7.16}$$

which are

$$\lambda = v \pm c_s. \tag{7.17}$$

For small density perturbations, $c_s^2 = \gamma(p_0/\rho_0^\gamma)\rho^{\gamma-1} \approx \gamma p_0/\rho_0$, which gives the usual definition of the (small-amplitude) sound speed $c_s = \sqrt{\gamma p_0/\rho_0}$.

The fact that the eigenvalues are real is a demonstration that the system of equations is *hyperbolic*. The eigenvalues indicate the speed of propagation of disturbances. In this fluid they propagate at the speed of sound measured in the rest-frame of the fluid.

7.3 Finite differences and stability

Now let's consider possible finite-difference representations of the equations. We notice that the simplest time differences give the time derivative $(\mathbf{u}_j^{(n+1)} - \mathbf{u}_j^{(n)})/\Delta t$ effectively at time $n + 1/2$ but position j, and the simplest space difference gives a derivative $(\mathbf{f}_{j+1}^{(n)} - \mathbf{f}_j^{(n)})/\Delta x$ at position $j + 1/2$. These expressions don't line up with one another so if we use them we'll certainly have only first-order accuracy. See Fig. 7.3. We could try to fix that by taking centered derivatives; but it turns out that may make things worse. The scheme may become unstable. But how do we analyse stability for this fluid? We have multiple coupled dependent variables. How do we deal with that? The answer, in summary, is that we find the combinations of dependent variables that behave in a way that is approximately uncoupled from the other combinations of dependent variables – in other words, the characteristic *modes* of the system. Then we analyse the Von Neumann stability of those modes separately.

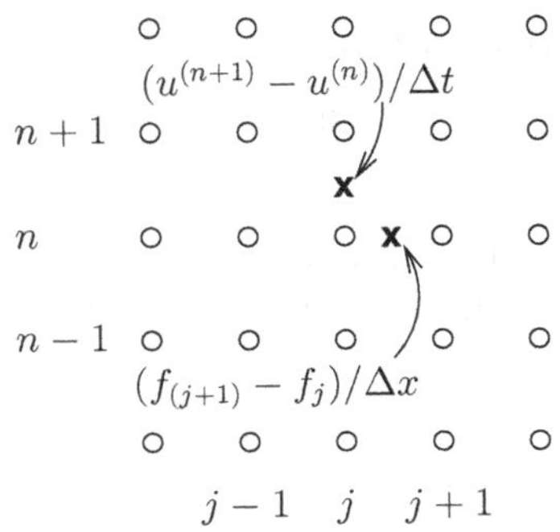

$(u^{(n+1)} - u^{(n)})/\Delta t$

$n+1$

n

$n-1$

$(f_{(j+1)} - f_j)/\Delta x$

$j-1 \quad j \quad j+1$

Figure 7.3 Derivatives in time (n) and space (j) implemented as finite differences give rise to values at the half-mesh points **x**.

If the Jacobian matrix is independent of position, then it is possible to change the dependent variables to new combinations of variables, each of which is *uncoupled* from the other combinations.[1] The new combination to use consists of the *eigenvectors* of the matrix **J**.

Let's illustrate this with our fluid. Consider the eigenvalues $\lambda = v \pm c_s$. For each eigenvalue, the eigenvector, which is the solution of the homogeneous equation $(\mathbf{J} - \lambda\mathbf{I})\mathbf{e} = 0$, must give zero when multiplied by any of the rows of the combined matrix $(\mathbf{J} - \lambda\mathbf{I})$. Using the top row, which becomes $(-[v \pm c_s], 1)$ we find that the eigenvector is proportional to

$$\mathbf{e}_\pm = \begin{pmatrix} 1 \\ v \pm c_s \end{pmatrix}. \tag{7.18}$$

Now we can express any vector state as the sum of two coefficients $q_\pm$ times the two eigenvectors[2] $\mathbf{u} = q_+\mathbf{e}_+ + q_-\mathbf{e}_-$ or written out

$$\begin{pmatrix} \rho \\ \Gamma \end{pmatrix} = q_+ \begin{pmatrix} 1 \\ v + c_s \end{pmatrix} + q_- \begin{pmatrix} 1 \\ v - c_s \end{pmatrix}. \tag{7.19}$$

[The coefficient values are $q_\pm = [\rho(v \mp c_s) - \Gamma]/(\pm 2c_s)$ but we don't need to know that.] The quantities $q_\pm$ can be considered to be the coefficients of the new vector representation $\mathbf{q} = \begin{pmatrix} q_+ \\ q_- \end{pmatrix}$, by which the state can be expressed.

[1] If the Jacobian is position-dependent, then the modes are only approximately uncoupled, because our analysis is effectively assuming that **J** and $\partial/\partial x$ commute, which is not exact if **J** is spatially varying. The stability analysis is then local, and only approximate. But in any case Von Neumann stability analysis is only approximate in non-uniform cases.

[2] This is provided the eigenvectors are linearly independent.

Then the result of multiplying $\mathbf{u}$ by the Jacobian matrix can be written in terms of the new set of q-coefficients as follows:

$$\mathbf{J}\mathbf{u} = q_+ \mathbf{J}\mathbf{e}_+ + q_- \mathbf{J}\mathbf{e}_- = q_+ \lambda_+ \mathbf{e}_+ + q_- \lambda_- \mathbf{e}_- . \tag{7.20}$$

This shows that the vector of eigenvalue coefficients giving $\mathbf{J}\mathbf{u}$ is $\begin{pmatrix} q_+\lambda_+ \\ q_-\lambda_- \end{pmatrix}$. So in terms of the new q-representation

$$\bar{\mathbf{J}}\mathbf{q} = \begin{pmatrix} q_+\lambda_+ \\ q_-\lambda_- \end{pmatrix} = \begin{pmatrix} \lambda_+ & 0 \\ 0 & \lambda_- \end{pmatrix} \begin{pmatrix} q_+ \\ q_- \end{pmatrix}. \tag{7.21}$$

In this q-representation, the operator $\mathbf{J}$ is represented by a different matrix $\bar{\mathbf{J}}$, which is *diagonal* having coefficients equal to the eigenvalues. Consequently the equations governing the evolution of the coefficients $\mathbf{q}$ of the eigenvectors separates into two independent equations

$$\frac{\partial q_\pm}{\partial t} = -\lambda_\pm \frac{\partial q_\pm}{\partial x}, \tag{7.22}$$

in place of the previously coupled equations governing $\mathbf{u}$. This process is totally general and will work for vectors of any dimensionality, corresponding to any order differential equations. We can now analyse each scalar equation separately for stability. Recognize, though, that the eigenvalues are *not* necessarily uniform in space, therefore this separation of the equations applies really only locally. So the stability analysis we now do is an approximate local analysis, not a precise global analysis.

7.3.1 FTCS is unstable

A forward time, centered space difference scheme might spring to mind as a natural one, illustrated in Fig. 7.4. For stability analysis purposes, we can suppose that we are using the new representation (in other words u stands for each q which we can consider separately in scalar equations). But we don't actually do the transformation to that new representation when implementing the scheme (only when analysing its stability). The first time through we'll do things explicitly but then take short cuts, thereafter not bringing q in explicitly. We write out the difference equation as

$$\mathbf{u}_j^{(n+1)} - \mathbf{u}_j^{(n)} = -\frac{\Delta t}{2\Delta x}\left(\mathbf{f}_{j+1}^{(n)} - \mathbf{f}_{j-1}^{(n)}\right) = -\frac{\Delta t}{2\Delta x}\mathbf{J}\left(\mathbf{u}_{j+1}^{(n)} - \mathbf{u}_{j-1}^{(n)}\right), \tag{7.23}$$

which in the new representation is

$$q_j^{(n+1)} - q_j^{(n)} = -\frac{\Delta t}{2\Delta x}\lambda\left(q_{j+1}^{(n)} - q_{j-1}^{(n)}\right) \tag{7.24}$$

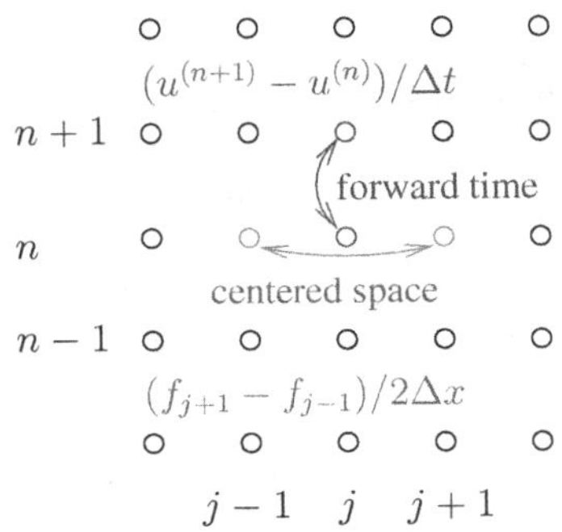

Figure 7.4 Forward derivative in time (n) and centered in space (j) (FTCS) finite differences give rise to an unstable scheme for hyperbolic problems.

and consider a single spatial Fourier mode of u and f and hence of q

$$q_j = q \exp(ik_x j\Delta x), \qquad f_j = f \exp(ik_x j\Delta x). \tag{7.25}$$

Substituting for the spatial dependence, the advancing equation becomes

$$q_j^{(n+1)} = [1 - \frac{\Delta t\lambda}{2\Delta x}(\mathrm{e}^{ik_x\Delta x} - \mathrm{e}^{-ik_x\Delta x})]q_j^{(n)} = [1 - i\frac{\Delta t\lambda}{\Delta x}\sin(k_x\Delta x)]q_j^{(n)}. \tag{7.26}$$

Now we see immediately that the temporal amplification factor is

$$A = 1 - i\frac{\Delta t\lambda}{\Delta x}\sin(k_x\Delta x). \tag{7.27}$$

Because the second term is imaginary, the magnitude of the amplification factor is always greater than 1, regardless of the (real) value of λ. All modes are unstable, growing with time! FTCS does not work for hyperbolic equations.

7.3.2 Lax–Friedrichs and the CFL condition

One tiny change works to stabilize the scheme. That is to replace $u_j^{(n)}$ on the left-hand side of eq. (7.23) with $(u_{j-1}^{(n)} + u_{j+1}^{(n)})/2$, as illustrated in Fig. 7.5. This is then called the Lax–Friedrichs method:

$$\mathbf{u}_j^{(n+1)} - (\mathbf{u}_{j-1}^{(n)} + \mathbf{u}_{j+1}^{(n)})/2 = -\frac{\Delta t}{2\Delta x}(\mathbf{f}_{j+1}^{(n)} - \mathbf{f}_{j-1}^{(n)}) = -\frac{\Delta t}{2\Delta x}\mathbf{J}(\mathbf{u}_{j+1}^{(n)} - \mathbf{u}_{j-1}^{(n)}). \tag{7.28}$$

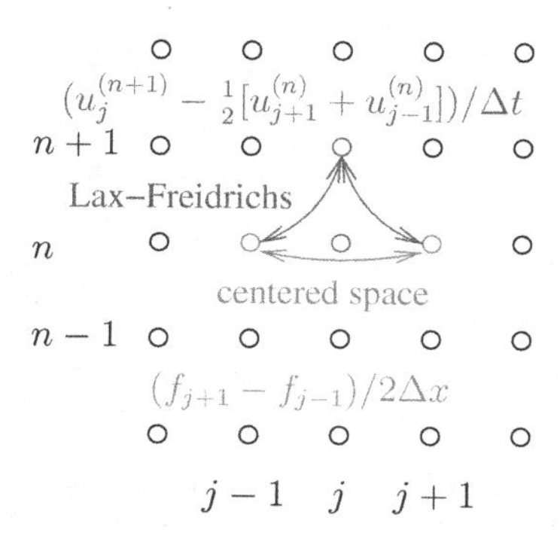

Figure 7.5 Forward derivative in time (n) but from the mean of the adjacent points and centered in space (j) is the Lax–Friedrichs finite-difference scheme, which is stable provided $\Delta t \leq \Delta x/|\lambda|$.

The student should verify that replacing $\mathbf{J}$ with λ for a scalar version of these equations, the resulting amplification factor is

$$A = \cos(k_x \Delta x) - i\frac{\Delta t \lambda}{\Delta x}\sin(k_x \Delta x). \tag{7.29}$$

As k_x varies, this is the equation of an ellipse in the complex plane. For stability, this ellipse must be entirely inside the unit circle, which requires the imaginary coefficient's magnitude to be less than or equal to 1

$$\Delta t \leq \Delta x/|\lambda|. \tag{7.30}$$

For our fluid example this is $\Delta t \leq \Delta x/|v \pm c_s|$. Equation (7.30) is called the Courant–Friedrichs–Lewy (CFL) condition. It applies to essentially all explicit schemes for hyperbolic equations. It says that Δt must be less than the time it takes for influence to propagate at the characteristic speed(s) (given by the eigenvalues of $\mathbf{J}$) from the prior adjacent nodes. If it were greater, then influence from other nodes, not taken into account in the difference scheme, would influence the solution.

7.3.3 Lax–Wendroff achieves second-order accuracy

The low-order errors of the Lax–Friedrichs scheme make it of little practical value. It has a substantial level of spurious *numerical diffusion* that damps out perturbations that should not be damped. For example, the simple fluid we've used to illustrate the issues has no physical dissipation, yet for some modes Lax–Friedrichs gives $|A|$ substantially less than one. They are damped.

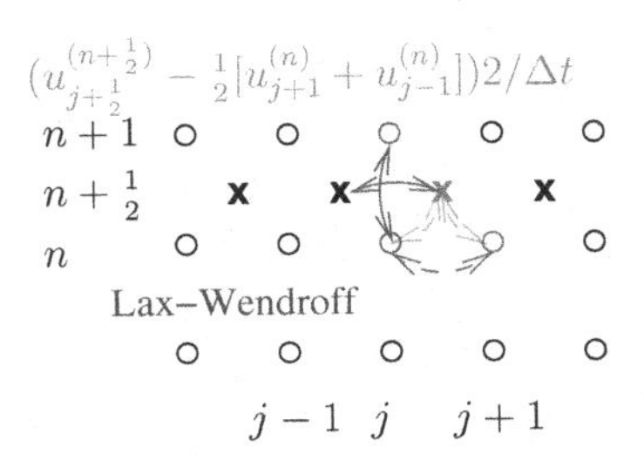

Figure 7.6 The Lax–Wendroff two-step scheme first (dashed lines) generates u and hence f values at the half-timestep $n + 1/2$, by a Lax–Friedrichs advance to (**X**). Then it uses a centered time, centered space full-step advance based upon $\mathbf{f}^{(n+1/2)}$, from the $\mathbf{u}^{(n)}$.

A better scheme, which is second order in time and still stable, is the Lax–Wendroff scheme. The advance is implemented in two steps:

$$\mathbf{u}_{j+1/2}^{(n+1/2)} = \frac{1}{2}\left(\mathbf{u}_{j+1}^{(n)} + \mathbf{u}_j^{(n)}\right) - \frac{\Delta t}{2\Delta x}\left(\mathbf{f}_{j+1}^{(n)} - \mathbf{f}_j^{(n)}\right) \tag{7.31}$$

$$\mathbf{u}_j^{(n+1)} = \mathbf{u}_j^{(n)} - \frac{\Delta t}{\Delta x}\left(\mathbf{f}_{j+1/2}^{(n+1/2)} - \mathbf{f}_{j-1/2}^{(n+1/2)}\right). \tag{7.32}$$

Fig. 7.6 shows the schematic. The first step is like a Lax–Friedrichs half-step to the half-way positions. Then the fluxes are evaluated again, at the half-step times and positions using the $\mathbf{u}^{(n+1/2)}$ values, to find the $\mathbf{f}^{(n+1/2)}$. Those are used in the second step to advance all the way from n to $n + 1$ in a properly centered manner. The amplification factor for the combined step can be shown to be

$$A = 1 - i\frac{\Delta t\lambda}{\Delta x}\sin(k_x\Delta x) + \left(\frac{\Delta t\lambda}{\Delta x}\right)^2[\cos(k_x\Delta x) - 1], \tag{7.33}$$

which gives stability if $\Delta t\lambda/\Delta x \leq 1$: the CFL condition, the same as before.

There are several other schemes in regular use for solving first-order hyperbolic problems to second-order accuracy. They practically all use multi-step approaches like the Lax-Wendroff method.

Worked example. Stability of Lax–Wendroff scheme

Derive the amplification factor for the Lax–Wendroff scheme and verify the stability condition $\Delta t\lambda/\Delta x \leq 1$.

Start with the formula for the first time half-step, eq. (7.31). For stability analysis (but not in implementing an actual numerical scheme), approximate the Jacobian matrix locally as uniform, and substitute $\mathbf{f} = \mathbf{J}\mathbf{u}$ at all the required mesh positions, deriving

$$\begin{aligned}\mathbf{u}_{j+1/2}^{(n+1/2)} &= \tfrac{1}{2}\left(\mathbf{u}_{j+1}^{(n)} + \mathbf{u}_j^{(n)}\right) - \tfrac{\Delta t}{2\Delta x}\mathbf{J}\left(\mathbf{u}_{j+1}^{(n)} - \mathbf{u}_j^{(n)}\right)\\ &= \tfrac{1}{2}\left[(\mathbf{I} - \tfrac{\Delta t}{\Delta x}\mathbf{J})\mathbf{u}_{j+1}^{(n)} + (\mathbf{I} + \tfrac{\Delta t}{\Delta x}\mathbf{J})\mathbf{u}_j^{(n)}\right].\end{aligned} \tag{7.34}$$

Similarly, the second half-step can be written:

$$\mathbf{u}_j^{(n+1)} = \mathbf{u}_j^{(n)} - \tfrac{\Delta t}{\Delta x}\mathbf{J}\left(\mathbf{u}_{j+1/2}^{(n+1/2)} - \mathbf{u}_{j-1/2}^{(n+1/2)}\right). \tag{7.35}$$

Substitute for the half-step values from eq. (7.34) to find:

$$\begin{aligned}\mathbf{u}_j^{(n+1)} - \mathbf{u}_j^{(n)} &= -\tfrac{\Delta t}{2\Delta x}\mathbf{J}\left[(\mathbf{I} - \tfrac{\Delta t}{\Delta x}\mathbf{J})\mathbf{u}_{j+1}^{(n)} + (\mathbf{I} + \tfrac{\Delta t}{\Delta x}\mathbf{J})\mathbf{u}_j^{(n)}\right.\\ &\qquad\left. -(\mathbf{I} - \tfrac{\Delta t}{\Delta x}\mathbf{J})\mathbf{u}_j^{(n)} - (\mathbf{I} + \tfrac{\Delta t}{\Delta x}\mathbf{J})\mathbf{u}_{j-1}^{(n)}\right]\\ &= -\tfrac{\Delta t}{2\Delta x}\mathbf{J}\left[(\mathbf{u}_{j+1}^{(n)} - \mathbf{u}_{j-1}^{(n)}) - \tfrac{\Delta t}{\Delta x}\mathbf{J}(\mathbf{u}_{j+1}^{(n)} - 2\mathbf{u}_j^{(n)} + \mathbf{u}_{j-1}^{(n)})\right].\end{aligned} \tag{7.36}$$

Now we consider an eigenmode of $\mathbf{J}$, so we can substitute the eigenvalue λ for $\mathbf{J}$, everywhere in the above expression. And we consider a spatial Fourier mode, for which $\mathbf{u}_j \propto e^{ik_x j\Delta x}$. The equation can then be written

$$\mathbf{u}_j^{(n+1)} - \mathbf{u}_j^{(n)} = -\tfrac{\Delta t}{2\Delta x}\lambda\left[(e^{ik_x\Delta x} - e^{-ik_x\Delta x}) + \tfrac{\Delta t}{\Delta x}\lambda(e^{ik_x\Delta x} - 2 + e^{-ik_x\Delta x})\right]\mathbf{u}_j^{(n)}, \tag{7.37}$$

or in other words

$$\mathbf{u}_j^{(n+1)} = \left\{1 - \tfrac{\Delta t\lambda}{\Delta x} i\sin(k_x\Delta x) + \left(\tfrac{\Delta t\lambda}{\Delta x}\right)^2[\cos(k_x\Delta x) - 1]\right\}\mathbf{u}_j^{(n)}. \tag{7.38}$$

The coefficient of $\mathbf{u}_j^{(n)}$ is the amplification factor A. Its squared absolute value is

$$\begin{aligned}|A|^2 &= \left\{1 + \left(\tfrac{\Delta t\lambda}{\Delta x}\right)^2[\cos(k_x\Delta x) - 1]\right\}^2 + \left\{\tfrac{\Delta t\lambda}{\Delta x}\sin(k_x\Delta x)\right\}^2\\ &= 1 + \left(\tfrac{\Delta t\lambda}{\Delta x}\right)^2[2\cos(k_x\Delta x) - 2 + \sin^2(k_x\Delta x)]\\ &\quad + \left(\tfrac{\Delta t\lambda}{\Delta x}\right)^4[\cos(k_x\Delta x) - 1]^2\\ &= 1 + \left[-\left(\tfrac{\Delta t\lambda}{\Delta x}\right)^2 + \left(\tfrac{\Delta t\lambda}{\Delta x}\right)^4\right][\cos(k_x\Delta x) - 1]^2.\end{aligned} \tag{7.39}$$

Thus $|A|^2 \le 1$ provided $\frac{\Delta t\lambda}{\Delta x} \le 1$, which is the stability criterion.

Worked example. Three-dimensional fluids

Formulate a finite-difference representation of the hyperbolic equations for a source-free, inviscid, isotropic fluid in three dimensions plus time, when the equation of state is $p = p(\rho)$. Assume the eigenvalue of the Jacobian of the linearized system (perturbation propagation speed) is known, λ, and that the eigenmode is longitudinal; deduce the condition governing the stable explicit timestep for centered spatial differences on a uniform cartesian grid spaced unequally in the different axis directions.

We use the density ρ and the flux density $\mathbf{\Gamma}$ as the elements of the state vector $\mathbf{u}$. In three dimensions, a vector quantity like $\mathbf{\Gamma}$ has three components, Γ_α, $\alpha = 1, 2, 3$. So the state vector has a total of four.

$$\mathbf{u} = \begin{pmatrix} \rho \\ \mathbf{\Gamma} \end{pmatrix} = \begin{pmatrix} \rho \\ \Gamma_1 \\ \Gamma_2 \\ \Gamma_3 \end{pmatrix} \tag{7.40}$$

The continuity 7.1 and momentum 7.5 equations are written with the time and space differentials separated on the left- and the right-hand sides, and we replace $\rho\boldsymbol{v}$ everywhere with $\mathbf{\Gamma}$.

$$\frac{\partial \rho}{\partial t} = -\nabla.(\rho\boldsymbol{v}) = -\nabla.\mathbf{\Gamma}, \tag{7.41}$$

$$\frac{\partial \mathbf{\Gamma}}{\partial t} = -\nabla.(\rho\boldsymbol{v}\boldsymbol{v}) - \nabla p = -\nabla.(\mathbf{\Gamma}\mathbf{\Gamma}/\rho - \mathbf{I}p). \tag{7.42}$$

In three dimensions, these are four scalar equations in total. The combined state-space form is

$$\frac{\partial \mathbf{u}}{\partial t} = -\nabla.\mathbf{f}, \tag{7.43}$$

where $\nabla. = \sum_\alpha \hat{\boldsymbol{x}}_\alpha.\frac{\partial}{\partial x_\alpha}$, the spatial-three-vector divergence, operates separately on each of the four (3-vector) entries of the state-space column-vector

$$\mathbf{f} = \begin{pmatrix} \mathbf{\Gamma} \\ \mathbf{\Gamma}\Gamma_1/\rho + p\hat{\boldsymbol{x}}_1 \\ \mathbf{\Gamma}\Gamma_2/\rho + p\hat{\boldsymbol{x}}_2 \\ \mathbf{\Gamma}\Gamma_3/\rho + p\hat{\boldsymbol{x}}_3 \end{pmatrix} = \begin{pmatrix} \mathbf{\Gamma} \\ \mathbf{\Gamma}\mathbf{\Gamma}/\rho + p\mathbf{I} \end{pmatrix}. \tag{7.44}$$

The spatial discrete difference scheme may be written in terms of three cartesian indices i, j, k of the mesh, as

$$\nabla.\mathbf{f} = \frac{\hat{\boldsymbol{x}}_1}{2\Delta x_1}.\left(\mathbf{f}^{(n)}_{(i+1)jk} - \mathbf{f}^{(n)}_{(i-1)jk}\right) + \frac{\hat{\boldsymbol{x}}_2}{2\Delta x_2}.\left(\mathbf{f}^{(n)}_{i(j+1)k} - \mathbf{f}^{(n)}_{i(j-1)k}\right)$$
$$+ \frac{\hat{\boldsymbol{x}}_3}{2\Delta x_3}.\left(\mathbf{f}^{(n)}_{ij(k+1)} - \mathbf{f}^{(n)}_{ij(k-1)}\right). \tag{7.45}$$

We are told that the eigenvalue of the state system is λ and the eigenmode is longitudinal.[3] Therefore, for a plane wave proportional to $\exp(i\boldsymbol{k}.\boldsymbol{x})$ that is an eigenmode of the state, each state-component of $\mathbf{f}$ is oriented in the spatial direction $\boldsymbol{k}$. Write the unit vector $\hat{\boldsymbol{k}} = (\hat{k}_1, \hat{k}_3, \hat{k}_3)$, and $\boldsymbol{k} = k\hat{\boldsymbol{k}}$. Then for this plane wave we can replace each $\hat{\boldsymbol{x}}_\alpha.\mathbf{f}$ with $\lambda\hat{k}_\alpha\mathbf{u}$ to obtain

$$\nabla.\mathbf{f} = \frac{1}{2}\left[\frac{\lambda\hat{k}_1}{\Delta x_1}\left(\mathbf{u}^{(n)}_{(i+1)jk} - \mathbf{u}^{(n)}_{(i-1)jk}\right) - \frac{\lambda\hat{k}_2}{\Delta x_2}\left(\mathbf{u}^{(n)}_{i(j+1)k} - \mathbf{u}^{(n)}_{i(j-1)k}\right)\right. \tag{7.46}$$
$$\left. - \frac{\lambda\hat{k}_3}{\Delta x_3}\left(\mathbf{u}^{(n)}_{ij(k+1)} - \mathbf{u}^{(n)}_{ij(k-1)}\right)\right].$$

Substituting for the variation of $\mathbf{u}$ with spatial index, e.g. $\left(\mathbf{u}^{(n)}_{ij(k+1)} - \mathbf{u}^{(n)}_{ij(k-1)}\right)$ $= \exp(ik_3\Delta x_3) - \exp(-ik_3\Delta x_3) = 2i\sin(k_3\Delta x_3)$, this form reduces the finite-difference equations to

$$\mathbf{u}^{(n+1)}_{ijk} - \mathbf{u}^{(n)}_s = -\Delta t\nabla.\mathbf{f} = -\sum_\alpha \frac{i\Delta t\lambda\hat{k}_\alpha}{\Delta x_\alpha}\sin(k\hat{k}_\alpha\Delta x_\alpha)\mathbf{u}_{ijk}, \tag{7.47}$$

where $\mathbf{u}^{(n)}_s$ denotes the mesh expression used for the current time (n). For example, a Lax–Friedrichs choice

$$\mathbf{u}^{(n)}_s = \frac{1}{6}\left(\mathbf{u}^{(n)}_{(i-1)jk} + \mathbf{u}^{(n)}_{(i+1)jk} + \mathbf{u}^{(n)}_{i(j-1)k} + \mathbf{u}^{(n)}_{i(j+1)k} + \mathbf{u}^{(n)}_{ij(k-1)} + \mathbf{u}^{(n)}_{ij(k+1)}\right)$$

leads to an amplification factor

$$A = \sum_\alpha\left[\frac{1}{3}\cos(k\hat{k}_\alpha\Delta x_\alpha) - \frac{i\Delta t\lambda\hat{k}_\alpha}{\Delta x_\alpha}\sin(k\hat{k}_\alpha\Delta x_\alpha)\right]. \tag{7.48}$$

We require $|A|^2 \leq 1$ for all modes to avoid instability. The worst case for stability occurs when all $\hat{k}_\alpha\Delta x_\alpha$ have the same value, which we'll denote $\Delta = (\sum_\alpha 1/\Delta x_\alpha^2)^{-1/2}$. Avoiding instability in this case requires that

$$\Delta t\lambda\sum_\alpha\frac{\hat{k}_\alpha^2}{\hat{k}_\alpha\Delta x_\alpha} = \frac{\Delta t\lambda}{\Delta} \leq 1. \tag{7.49}$$

Notice, by considering $k\Delta = \pi/2$, that the criterion must be satisfied for stability, regardless of the precise form chosen for $\mathbf{u}_s$, so long as that form

[3] Although this can be proved, it is complicated.

is symmetric in each coordinate direction, and hence gives rise to a *real* contribution to A. The criterion is thus universally *necessary* for any symmetric centered spatial difference scheme, when time differences are explicit, but it is not always *sufficient*.

When all the Δx_α are equal, then $\Delta = \Delta x/\sqrt{3}$, and the CFL condition for stability when v is small (so $\lambda = c_s$) is

$$\Delta t \leq \frac{\Delta x}{c_s\sqrt{3}}.$$

If, by contrast, for some direction β, Δx_β is much smaller than the other two grid spacings, then $\Delta \approx \Delta x_\beta$ and stability requires $\Delta t \leq \Delta x_\beta / c_s$.

Exercise 7. Fluids and hyperbolic equations

1. Prove eq. (7.29), the amplification factor for the Lax–Friedrichs scheme.
2. Consider an isothermal gas in one dimension. It obeys the equations

continuity: $$\frac{\partial \rho}{\partial t} + \frac{\partial}{\partial x}(\rho v) = 0,$$

momentum: $$\frac{\partial}{\partial t}(\rho v) + \frac{\partial}{\partial x}(\rho v^2) = -\frac{\partial}{\partial x}p,$$

state: $$p = \rho(kT/m),$$

with kT/m simply a constant equal to the ratio of the temperature in energy units to the gas molecule mass m.

(a) Convert this into the form of a state and flux vector equation

$$\frac{\partial \mathbf{u}}{\partial t} = -\frac{\partial \mathbf{f}}{\partial x}.$$

where

$$\mathbf{u} = \begin{pmatrix} \rho \\ \Gamma \end{pmatrix}$$

is the state vector ($\Gamma = \rho v$) and you should give the flux vector $\mathbf{f}$.

(b) Calculate the Jacobian matrix $\mathbf{J} = \partial \mathbf{f}/\partial \mathbf{u}$.

(c) Find its eigenvalues.

3. Find the finite difference form and CFL stability when the linearized eigenmode is longitudinal with eigenvalue λ, for the Lax-Wendroff scheme in two space-dimensions.

8
Boltzmann's equation and its solution

So far in our discussion of multidimensional problems we have been focussing on continuum fluids governed by partial differential equations. Despite the fact that treating fluids as continua seems entirely natural, and gives remarkably accurate representation in many cases, we know that fluids in nature are not continuous. They are made up of individual molecules. A continuum representation is expected to work well only when the molecules experience collisions on a time and space scale much shorter than those of interest to our situation. By contrast, when the collision mean-free-path is either an important part of the problem, as it is, for example, when calculating the viscosity of a fluid, or when the collision mean-free-path (or time) is long compared with the typical scales of the problem, as it is for very dilute gases and for many plasmas, a fluid treatment cannot cope. We then need to represent the discrete molecular nature of the substance as well as its collective behavior.

Even so, it is unrealistic in most problems to suppose that we can follow the detailed dynamics of each individual molecule. There are $p/kT = 10^5(\mathrm{Pa})/[1.38 \times 10^{-23}(\mathrm{J/K}) \times 273(\mathrm{K})] = 2.65 \times 10^{25}$ molecules, for example, in a cubic meter of gas at atmospheric pressure and 0°C temperature (STP). Even computers of the distant future are not going to track every particle in such an assembly. Instead, a statistical description is used. The treatment is common to many different types of particles. The particles under consideration might be neutrons in a fission reactor, neutral molecules of a gas, electrons of a plasma, and so on.

8.1 The distribution function

Consider a volume element, small compared with the size of the problem but still large enough to contain very many particles. The element is cuboidal

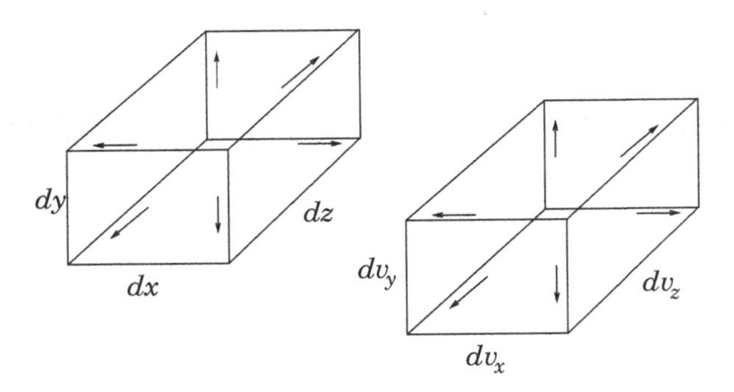

Figure 8.1 The phase-space element is six-dimensional and selects particles that lie in a space element d^3x and simultaneously in a velocity element d^3v.

$d^3x = dx.dy.dz$ with sides $\boldsymbol{dx}$, see Fig. 8.1. It is located at the position $\boldsymbol{x}$. We want a sufficient description of the average properties of the particles in this element.

We use a statistical description originally invented by James Clerk Maxwell called the "distribution function." The distribution function is a quantity $f(\boldsymbol{v}, \boldsymbol{x}, t)$ that is a function of velocity $\boldsymbol{v}$, position $\boldsymbol{x}$, and time t. The distribution function is defined by considering an element in *velocity-space* $d^3v = dv_x.dv_y.dv_z$ with sides $\boldsymbol{dv}$ located at velocity $\boldsymbol{v}$. Any particle whose velocity components lie simultaneously in the ranges $v_x \to v_x + dv_x$, $v_y \to v_y + dv_y$, $v_z \to v_z + dv_z$, is inside that velocity element.

Then at any time t the distribution function $f(\boldsymbol{v}, \boldsymbol{x}, t)$ is such that the number of particles in the spatial element d^3x that have velocities in the velocity element d^3v is

$$f(\boldsymbol{v}, \boldsymbol{x}, t) d^3v \, d^3x. \tag{8.1}$$

The distribution function is therefore the density of particles in the six-dimensional "phase-space" combining velocity and space. Its utility arises from the presumption that because of the enormous number of particles in the problem we can let the velocity and spatial elements, that is the phase-space element $d^3v \, d^3x$, become almost infinitesimally small and yet still have a large number of particles in it. With a large number of particles, statistical descriptions make sense. In particular, it makes sense to think of f as a kind of continuous fluid in six dimensions. Obviously if the phase-space element is shrunk down to a sufficiently small size, then eventually there will be very few particles in it. The discreteness of the particles becomes visible and eventually there are either one or no particles in each tiny volume. But if we can shrink the element enough that it is small compared with the smallest scales in the problem while it still contains a large number of particles, then we have a sufficient statistical description if we know f everywhere, but we don't know the coordinates of each individual particle. The most famous of all such distribution functions is the Maxwellian, which is

$$f(\boldsymbol{v},\boldsymbol{x},t) = n(\boldsymbol{x},t)\left(\frac{m}{2\pi kT}\right)^{3/2}\exp\left(-\frac{mv^2}{2kT}\right). \tag{8.2}$$

Here m is the mass of the particles, T their temperature, and k Boltzmann's constant. The squared velocity $v^2 = \boldsymbol{v}.\boldsymbol{v} = v_x^2 + v_y^2 + v_z^2$ appearing in the exponential makes the distributions in the different coordinate directions separable

$$\exp\left(-\frac{mv^2}{2kT}\right) = \exp\left(-\frac{mv_x^2}{2kT}\right)\exp\left(-\frac{mv_y^2}{2kT}\right)\exp\left(-\frac{mv_z^2}{2kT}\right). \tag{8.3}$$

See Fig. 8.2. The factor $(m/2\pi kT)^{3/2}$ normalizes the distributions in the three velocity dimensions. It is equal to the inverse of the integral of eq. (8.3) over all velocities. Therefore the leading term $n(\boldsymbol{x}, t)$ is just the density in space (not phase-space). It might vary with position or time. The Maxwellian distribution is what occurs in *thermodynamic equilibrium* when there are no substantial effects driving the velocity distribution away from its natural form. But there

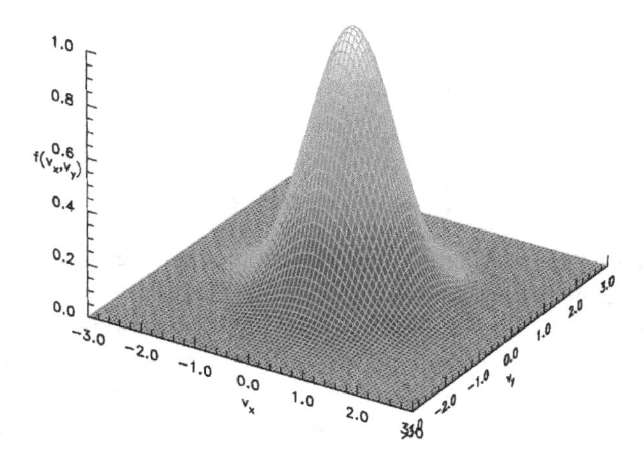

Figure 8.2 A Maxwellian distribution function in two dimensions displayed as a perspective view of the surface $f(v_x, v_y)$, with velocities (v_x, v_y) normalized to the thermal velocity $\sqrt{2T/m}$. This can be considered to be the proportional to the distribution at a fixed value of v_z, since the Maxwellian is separable.

are many important situations where non-thermal, that is non-Maxwellian, distributions arise.

The distribution function directly determines the mean flow velocity, and for particles whose internal energy is unimportant, the energy density. The particle flux density, which is the fluid velocity times the fluid density, is

$$\mathbf{\Gamma} = n\boldsymbol{v} = \int \boldsymbol{v} f(\boldsymbol{v}, \boldsymbol{x}, t) d^3 v. \tag{8.4}$$

The kinetic energy density, which for a stationary fluid can be considered the density times 3/2 times the temperature is

$$\mathcal{E} = \frac{3}{2} nkT = \int \frac{1}{2} m v^2 f d^3 v. \tag{8.5}$$

When the distribution in a specific coordinate direction does not concern us, perhaps because it is known, or because by symmetry it is unimportant, we often reduce the number of dimensions which we track. For example, often we might address only the x-direction velocity v_x. In that case, we use a one-dimensional distribution function,

$$f_x(v_x) = \int f(\boldsymbol{v}) dv_y dv_z, \tag{8.6}$$

that is the *integral* of the full three-dimensional distribution over the ignorable velocity coordinates. In effect, $f_x(v_x)$ picks out a particular v_x but includes all possible v_y and v_z. So the number of particles in the velocity element dv_x is $f_x(v_x)dv_x$.

The distribution function first arose in connection with the kinetic theory of gases. Its use is therefore often referred to as "kinetic theory."

8.2 Conservation of particles in phase-space

Boltzmann's equation governs the conservation of particles; not just in space, which is the continuity equation (4.5), but in phase-space. When solved, it tells us what the distribution function actually is. Its derivation is mathematically very much like the derivation of the fluid continuity equation. The main complication is that one needs to think in six dimensions! We'll usually illustrate this thinking in a (two-dimensional) diagram, using just one space (x as the abscissa) and one velocity (v as the ordinate) dimension. See Fig. 8.3. As time passes, particles move in phase-space. The rate of change of x is the velocity $dx/dt = v$. The rate of change of velocity v is acceleration $dv/dt = a$. Generally, acceleration arises from force (per particle) divided by particle

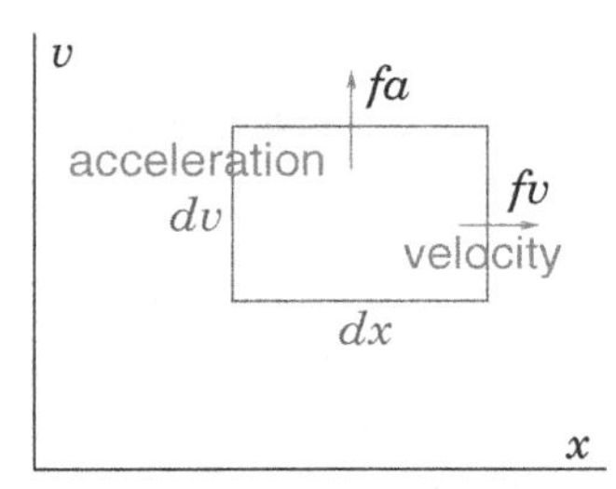

Figure 8.3 In phase-space, velocity v carries a particle in the x-direction, acceleration carries it in the v-direction. Particle flux out of an element $dvdx$ arises from the divergence of the fluxes fv and fa in the respective directions.

mass. The force might be gravity, or (for charged particles) an electric or a magnetic field. An individual particle thus moves through the phase-space (in the xv plane of our diagram). If we therefore consider some phase-space volume we can write the conservation of particles within it just as we did for the fluid continuity equation as

$$\frac{\partial f}{\partial t} + \nabla_{ps}.(f\boldsymbol{v}_{ps}) = \frac{\partial f}{\partial t} + \frac{\partial}{\partial \boldsymbol{x}}.(f\boldsymbol{v}) + \frac{\partial}{\partial \boldsymbol{v}}.(f\boldsymbol{a}) = S. \tag{8.7}$$

Here $\boldsymbol{v}_{ps}$ is the "phase-space velocity," a six-dimensional vector consisting of the combination of the velocity in space and the acceleration. And ∇_{ps} is the gradient operator in phase-space, likewise a six-dimensional vector:

$$\boldsymbol{v}_{ps} = \begin{pmatrix} \boldsymbol{v} \\ \boldsymbol{a} \end{pmatrix} = \begin{pmatrix} v_x \\ v_y \\ v_z \\ a_x \\ a_y \\ a_z \end{pmatrix}, \text{ and } \nabla_{ps} = \begin{pmatrix} \nabla \\ \nabla_v \end{pmatrix} = \begin{pmatrix} \dfrac{\partial}{\partial \boldsymbol{x}} \\ \dfrac{\partial}{\partial \boldsymbol{v}} \end{pmatrix} = \begin{pmatrix} \partial/\partial x \\ \partial/\partial y \\ \partial/\partial z \\ \partial/\partial v_x \\ \partial/\partial v_y \\ \partial/\partial v_z \end{pmatrix}. \tag{8.8}$$

The notation most usually used is to write out the space and velocity parts of the derivatives separately:

$$\frac{\partial}{\partial \boldsymbol{x}}.(f\boldsymbol{v}) = \frac{\partial(fv_x)}{\partial x} + \frac{\partial(fv_y)}{\partial y} + \frac{\partial(fv_z)}{\partial z} \tag{8.9}$$

and

$$\frac{\partial}{\partial \boldsymbol{v}}.(f\boldsymbol{a}) = \frac{\partial(fa_x)}{\partial v_x} + \frac{\partial(fa_y)}{\partial v_y} + \frac{\partial(fa_z)}{\partial v_z}. \tag{8.10}$$

This helps us remember we are dealing with phase-space. Equation (8.7) expresses the fact that the rate of change of the number of particles in a phase-space element is equal to the rate at which they are flowing inward across its boundary plus the combined source rate inside the element S (all per unit volume). Particles flow across the boundary either by moving in space across the boundary of d^3x or by accelerating (moving in velocity-space) across the boundary of d^3v.

A final simplification arises in eq. (8.7) because of what a partial derivative means. It means take the derivative *holding all the other phase-space coordinates constant.* In other words, the partial x-derivative holds y, z, v_x, v_y, v_z constant. The partial derivative of any v_j with respect to any x_k is therefore zero, which means that in the spatial divergence eq. (8.9) the velocity factors can be taken outside the spatial derivatives to write

$$\frac{\partial}{\partial \boldsymbol{x}}.(f\boldsymbol{v}) = \boldsymbol{v}.\frac{\partial f}{\partial \boldsymbol{x}} = v_x\frac{\partial f}{\partial x} + v_y\frac{\partial f}{\partial y} + v_z\frac{\partial f}{\partial z}. \tag{8.11}$$

That rearrangement is always possible. If the acceleration of a particle does not depend on its velocity (or depends on it in such a way that $\nabla_v.\boldsymbol{a} = 0$, which is the case for the Lorentz force) then we can do the same for the acceleration term.

Then we arrive at Boltzmann's equation

$$\frac{\partial f}{\partial t} + \boldsymbol{v}.\frac{\partial f}{\partial \boldsymbol{x}} + \boldsymbol{a}.\frac{\partial f}{\partial \boldsymbol{v}} = S = C. \tag{8.12}$$

The source term on the right-hand side of Boltzmann's equation contains not only literal creation or destruction of particles (e.g. by chemical or nuclear reactions), but also any instantaneous changes of velocity, in other words *collisions*. A collision that does not destroy or create a particle of the type we are tracking can nevertheless change its velocity abruptly.[1] This change in velocity immediately transports the particle from one velocity to another. The particle jumps to a different position in phase-space. That constitutes a "sink" at the old velocity and a "source" at the new velocity. Chemical or nuclear reactions also occur as a result of collisions, of course. Consequently, essentially all the phenomena that contribute to the Boltzmann equation's source (with the exception of spontaneous – e.g. radioactive – decay of the particles) are collisions; and the source term is usually called the "collision" term and written C instead of S.

The collision term $C(\boldsymbol{v}, \boldsymbol{x}, t)$ is the rate per unit phase-space-volume of generation (or removal if it is negative) of particles at position $\boldsymbol{x}$ having velocity $\boldsymbol{v}$. It naturally depends also upon the distribution function f itself. For

[1] There is considerable subtlety in the question "how abruptly." After all, any velocity change really takes a finite time and involves an acceleration. Why isn't it included in $\boldsymbol{a}$? For present purposes we'll just finesse this question by saying that the source contains any velocity changes that are not included in the acceleration.

example, the rate per unit volume at which collisions occur removing particles of a certain velocity is proportional to the number of such particles present in the first place.

8.3 Solving the hyperbolic Boltzmann equation

8.3.1 Integration along orbits

If we know the collision term C, as well as $\boldsymbol{a}$, then clearly Boltzmann's equation is a first-order linear partial differential equation (in seven total dimensions including time, or less if there are ignorable coordinates). Since it is first-order linear with a single scalar dependent variable,[2] f, it is hyperbolic. That means we may solve it as an initial-value problem.

The most natural way to do this is to follow particle trajectories in phase-space, which we will call particle orbits. Any individual particle moves in accordance with

$$\frac{d\boldsymbol{x}}{dt} = \boldsymbol{v}, \qquad \frac{d\boldsymbol{v}}{dt} = \boldsymbol{a}; \qquad \text{i.e.} \quad \frac{d}{dt}\begin{pmatrix}\boldsymbol{x}\\ \boldsymbol{v}\end{pmatrix} = \begin{pmatrix}\boldsymbol{v}\\ \boldsymbol{a}\end{pmatrix} \tag{8.13}$$

This is an ordinary differential equation, which we know how to solve (assuming $\boldsymbol{a}$ is known), starting from some initial position in phase-space $\boldsymbol{x}_0$, $\boldsymbol{v}_0$. But how does this orbit help us to solve Boltzmann's equation for the distribution function? It helps us because Boltzmann's equation is an equation for the rate of change of f moving along a particle phase-space orbit.

Suppose we catch a six-dimensional ride on one of the particles and move with it, looking out into the phase-space near to us and measuring the particle density there; watching it change as time goes by. The rate of change of the distribution function will be precisely that given by the left-hand side of Boltzmann's equation. See Fig. 8.4 for a visualization. First, let's convince ourselves that is true. If, during our ride, we measure the difference in f between times different by a small interval dt. The f will be different because (1) it may have intrinsic variation with time, resulting in a change $dt\,\partial f/\partial t$; (2) it may have variation with space so our motion has carried us a distance $dt\,\boldsymbol{v}$ to a place where f is different by $dt\,\boldsymbol{v}.\partial f/\partial \boldsymbol{x}$; or (3) it may have variation with velocity so our motion in velocity-space (acceleration) has carried us a velocity-"distance" $dt\,\boldsymbol{a}$ to where f is different by $dt\,\boldsymbol{a}.\partial f/\partial \boldsymbol{v}$. The total of these three, divided by dt, is the rate of change of f along the orbit. That's the left-hand side of eq. (8.12).

[2] (not multiple dependent variables that have to be arranged into a vector representation whose eigenvalue might not be real)

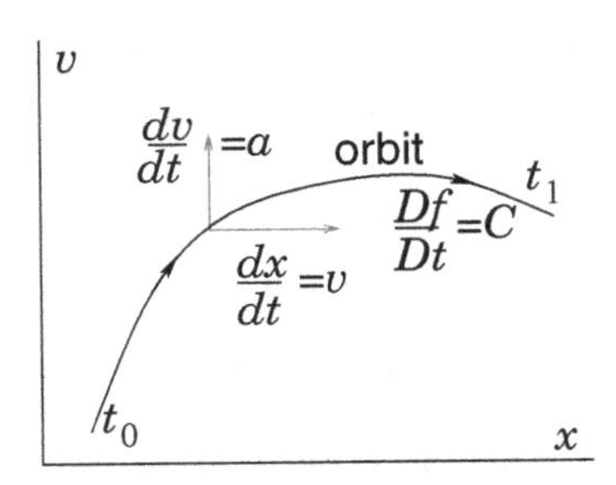

Figure 8.4 A phase-space orbit is determined by a first-order ordinary differential equation. The Boltzmann equation states that the rate of change of the distribution function along phase-space orbits is equal to the collision term.

Second, why is this identity with the total derivative the case? It is because the flow in phase-space is *divergenceless*, $\nabla_{ps}.\boldsymbol{v}_{ps} = 0$. Just as for an ordinary three-dimensional fluid for which $\nabla.(\rho\boldsymbol{v}) = \boldsymbol{v}.\nabla\rho$ if $\nabla.\boldsymbol{v} = 0$, implying $D\rho/Dt = S$, similarly for a dimensionless flow in phase-space. The phase-space flow is divergenceless if the acceleration has the requisite property $\nabla_v.\boldsymbol{a} = 0$. This can be interpreted as a statement that there is no dissipation.

Third, how does this identity help us? It reduces Boltzmann's equation to an ordinary differential equation along the orbits. Writing the total differential as D/Dt it becomes

$$\frac{\partial f}{\partial t} + \boldsymbol{v}.\frac{\partial f}{\partial \boldsymbol{x}} + \boldsymbol{a}.\frac{\partial f}{\partial \boldsymbol{v}} = \frac{Df}{Dt} = C. \tag{8.14}$$

We can integrate the second equality immediately to obtain:

$$f(\boldsymbol{v}_1, \boldsymbol{x}_1, t_1) - f(\boldsymbol{v}_0, \boldsymbol{x}_0, t_0) = \int_0^1 C dt. \tag{8.15}$$

The integral is along the orbit, whatever that might be in phase-space, from the initial position (0) to the final position (1). The final value of the distribution function, measured at the final values of velocity, position, and time, is equal to the initial value of the distribution function at the initial values of velocity, position, and time, plus the integral of the collision term along the orbit. The easiest case to deal with is if there are *no* collisions, $C = 0$. Then the initial and final distribution functions are equal in value. A fact we express by saying, in the absence of collisions, the distribution function is "constant along orbits."

It is vital to realize that constant along orbits generally does *not* mean that the distribution function is the same function of velocity at 1 as it was at 0. No, the orbital *velocity* has changed between those positions. So although f has the

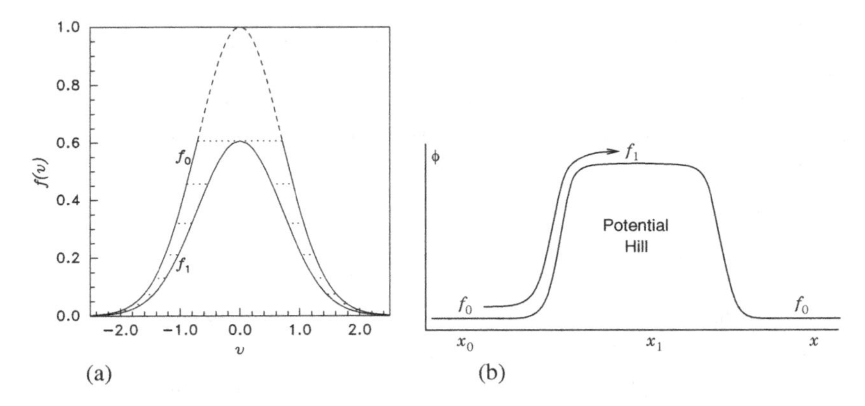

Figure 8.5 In the collisionless Boltzmann equation the distribution is constant along orbits. The distribution (a) is different at the top of a potential hill (b) because the speed on an orbit is smaller (conserving energy). The distribution values $f_0 = f(x_0)$ and $f_1 = f(x_1)$ are the same but at different velocities. Orbits have moved the distribution along the horizontal dotted lines in (a). The lowest velocity orbits of distribution f_0 (upper dashed part) can't reach the top of the hill where f_1 is, and do not contribute to it.

same *height*, that height does not occur at the same *velocity*. Fig. 8.5 illustrates this fact.

8.3.2 Orbits are characteristics

Every hyperbolic partial differential equation can be analysed in a manner equivalent to the integration along orbits. This approach is called the method of characteristics. The terminology "characteristic" is the general term for what we've called in the context of the Boltzmann equation an "orbit." Suppose we have a first-order linear equation, an advection equation with source

$$\mathbf{v}.\frac{\partial}{\partial \mathbf{x}}\psi = S, \tag{8.16}$$

in which the components of the N-dimensional vector $\mathbf{v}$ are simply known functions of ψ and the N-dimensional independent variables $\mathbf{x}$. Introduce a new parameter t, which is going to serve like time. (If the original equation already contained time as one of the independent variables, treat it just as one component of the vector $\mathbf{v}$ and use the new t as a parameterization.) Think of the $\mathbf{v}$ as velocities in N-dimensional space such that $d\mathbf{x}/dt = \mathbf{v}$. Remember, in the original formulation, those $\mathbf{v}$ were just the coefficients of the partial derivatives in the respective directions. What we've done is to address the question "what if the coefficients $\mathbf{v}$ were velocities?" in respect of the new t parameter we introduced. The answer is that starting from any point $\mathbf{x}$ we would move with the velocity $\mathbf{v}$ and thereby trace out an orbit in the $\mathbf{x}$-space. This orbit is what is called the characteristic of the differential equation. As we follow the characteristic, the equation we would be satisfying is

$$\mathbf{v}.\frac{\partial}{\partial \mathbf{x}}\psi = \frac{d\mathbf{x}}{dt}.\frac{\partial}{\partial \mathbf{x}}\psi = \sum_{j=1}^{N} \frac{dx_j}{dt}.\frac{\partial}{\partial x_j}\psi = \left.\frac{d\psi}{dt}\right|_{\text{orbit}} = S. \tag{8.17}$$

This equation is an ordinary differential equation along the characteristic and can be integrated as $\psi_1 - \psi_0 = \int_0^1 S dt$. This process is exactly what we did for Boltzmann's equation. The only difference is that Boltzmann's equation already contained time. Fortunately, the coefficient of $\partial/\partial t$ in Boltzmann's equation is 1. Therefore it was possible to choose the time-like parameter to be actual physical time, which we did. We could, however, have made a different choice, if we'd preferred. We also used notation familiar from fluid theory for the "convective" derivative D/Dt, but that is no different from d/dt along the orbit.

Higher-order scalar equations can be rendered into first-order, *vector* (multiple dependent variable) equations, as we saw before. If they are hyperbolic, then they have characteristics which correspond to the coefficients of the eigenvectors of the equations. Those eigenvectors must be real otherwise the presumption that there are real characteristics breaks down. The condition, therefore, for a system of vector equations to be hyperbolic is that they are diagonalizable with real eigenvectors.

8.4 Collision term

The importance of the Boltzmann equation's collision term depends upon the application. In some plasma and gravitational applications it can be completely neglected and ignored. Collisions are nothing. Then the equation of interest is called the Vlasov equation:

$$\frac{\partial f}{\partial t} + \boldsymbol{v}.\frac{\partial f}{\partial \boldsymbol{x}} + \boldsymbol{a}.\frac{\partial f}{\partial \boldsymbol{v}} = 0. \tag{8.18}$$

At the other extreme, when applied forces are negligible so $\boldsymbol{a} = 0$, and the problem is homogeneous and steady state $\partial/\partial \boldsymbol{x} = 0$, $\partial/\partial t = 0$, all that is left of the Boltzmann equation is $C = 0$. So then collisions are everything!

Also, the form of the collision term depends upon application. Especially, it depends upon whether the important collisions are with the same particles or with some particles of a different type that are described by a different distribution function.

8.4.1 Self-scattering

Self-scattering dominates, for example, a simple unreactive monatomic gas. There is just one species of particle. And elastic self-scattering is the only type of collision present. Then the integrals over all velocity-space of C, $\mathbf{v}C$, and v^2C are zero. That is a simple consequence of particle, momentum, and energy conservation. With self-scattering, though, other complications are severe. The rate at which collisions take place depends upon a product $f(\mathbf{v}_1)f(\mathbf{v}_2)$ of the distribution functions of the two colliding particles of different velocities (so it is non-linear). It is multiplied by the collision rate, which is the cross-section (a function of relative speed) times the relative speed $\sigma|\mathbf{v}_1 - \mathbf{v}_2|$. It is then integrated over the velocity $\mathbf{v}_2$ of the target particle (so Boltzmann's equation becomes an integro-differential equation). Generally, substantial approximation is necessary to make the collision term managable, even for numerical solution.

8.4.2 No self-scattering

If, however, the dominant interactions are with a different type of particle, then momentum and energy of the first species is not necessarily conserved. It can be transferred to the second species. But at least the collision term is linear in f, and if the initial velocity distribution of the second species is known or may be neglected, then substantial reduction of the need for integration can occur.

For example, as illustrated in Fig. 8.6, a form of collision term that approximately represents charge-exchange collisions at a fixed rate ν between singly charged ions (the species whose Boltzman equation we are trying to solve) and neutrals of the same element (the target species 2) is

$$C(f) = -\nu f(\mathbf{v}) + \nu f_2(\mathbf{v}). \tag{8.19}$$

This is sometimes called the BGK collision form. It represents depletion of the original ions at rate ν giving the term $-\nu f(\mathbf{v})$, and their direct replacement at the same rate by new ions. The newly born ions, before their collision, were neutrals. They retain the velocity distribution $f_2(\mathbf{v})$ they had before the collision, because the collision just transfers an electron from one to the other.

Another idealized example (Fig. 8.7) is when collisions are with heavy stationary targets (which therefore acquire negligible recoil energy), which happen to scatter equally, isotropically, in all directions. In a collision, a

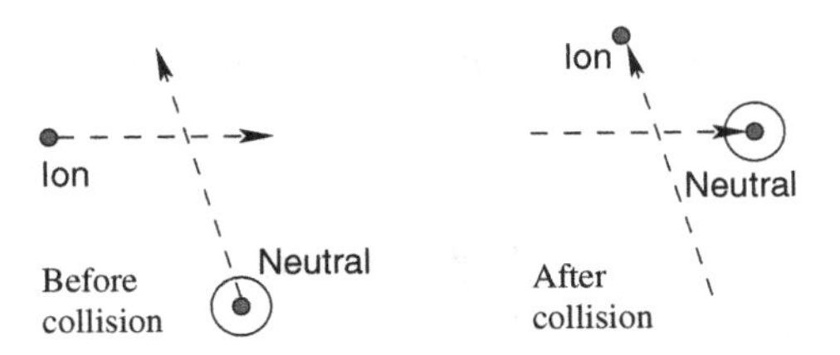

Figure 8.6 Charge exchange collisions, where an electron is transferred from a neutral to an ion, give rise to a simple collision term. If they occur at a constant rate, ν, eq. (8.19) applies.

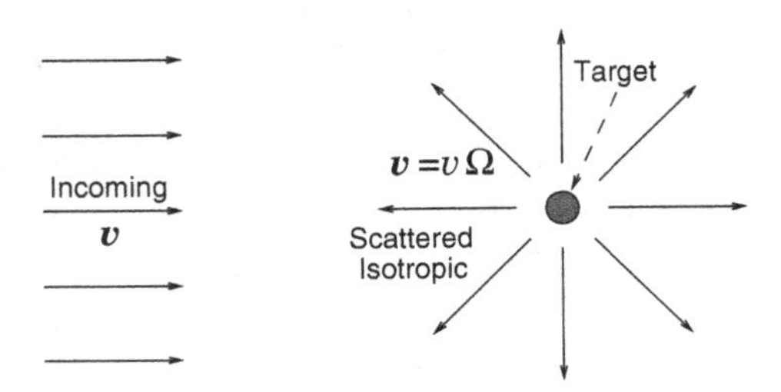

Figure 8.7 Isotropic scattering (an idealized approximation) gives particles emerging equally in all directions Ω. With heavy targets, v is not changed in magnitude, only in direction. Eq. (8.20) is the result.

particle just changes the direction of its velocity, not its magnitude. If the density of targets is n_2 and the collision cross-section is σ, then

$$C(f) = -n_2 \sigma v \left(f(\boldsymbol{v}) - \int f(\boldsymbol{v}) d^2\Omega / 4\pi \right). \tag{8.20}$$

Here $d^2\Omega = \sin\theta d\theta d\chi$ is the element of solid angle, and the integral is over the angular position (θ, χ) on the surface of the sphere in velocity-space at constant total velocity v. In other words, the second term is the average of the distribution function over all directions, at v. This type of collision scatters the velocity direction, thus tending to remove any anisotropy (variation with angles θ or χ.)

It should be noticed that in these examples where self-collisions can be ignored the collision term generally consists of two parts. The first is negative, the removal or "sink" rate of particles that collide with whatever targets happen to be present ($-n_2\sigma v f$ in eq. (8.20)). The second is positive, the "source" rate of particles from all mechanisms. When non-reactive gases are being treated, the source is only the re-emergence of particles from collisions. But in other situations, such as neutron transport in a reactor, generation of new particles from reactions or spontaneous emission from the target medium may be equally important.

For multiple target species j, the sink term is the sum of collisions with all target types. And this is often written in shorthand as $-\Sigma_t \times (vf)$, with

$$\Sigma_t = \sum_j n_j\sigma_j; \tag{8.21}$$

and referred to in reactor-physics literature as the "macroscopic cross-section." This terminology is unfortunate because the quantity Σ_t has units m^{-1} not m^2, and is an inverse attenuation-length, not a cross-section. When the targets are stationary, Σ_t is isotropic: the rate of collisions is independent of the direction of particle velocity. The source term, by contrast, is *not* usually isotropic because it includes the emergence of particles from pure scattering events. Scattering, even from stationary targets, usually partially retains any anisotropy in the distribution function itself. (The conditions of eq. (8.20) are a non-typical idealization.)

Worked example. Solving Vlasov's equation

Consider a steady-state situation, one dimensional in space and velocity, where acceleration arises only from a spatially varying potential energy of the form $\phi(x) = \phi_0 \exp(-x^2/w^2)$, so $a = -\frac{1}{m}d\phi/dx$, and collisions are negligible. If the distribution function at $|x| \to \infty$ is equal to $f_\infty(v) = \exp(-mv^2/2T)$, and $\phi_0 \geq 0$, find the distribution function $f(v, x)$ for all x and v. Can you solve this problem if $\phi_0 < 0$?

The steady collisionless one-dimensional Boltzmann (Vlasov) equation is

$$0 = \frac{Df}{Dt} = v.\frac{\partial f}{\partial x} + a.\frac{\partial f}{\partial v} \tag{8.22}$$

The equations of the orbit (the characteristics) are

$$\frac{dx}{dt} = v; \qquad \frac{dv}{dt} = a = -\frac{1}{m}\frac{d\phi}{dx}. \tag{8.23}$$

Multiplying the second of these by the first we find

$$v\frac{dv}{dt} + \frac{1}{m}\frac{d\phi}{dx}\frac{dx}{dt} = 0, \tag{8.24}$$

which may be immediately integrated to find

$$\frac{1}{2}mv^2 + \phi = const. \tag{8.25}$$

We have derived the conservation of energy, kinetic plus potential. The constant can be considered to be the kinetic energy at infinity, $mv_\infty^2/2$, where the potential (ϕ_∞) is zero.

For Vlasov's equation, the distribution function f is constant along the orbits. Therefore,

$$f(v,x) = f_\infty(v_\infty) = \exp(-mv_\infty^2/2T) = \exp(-[mv^2 + 2\phi(x)]/2T). \quad (8.26)$$

Substituting for $\phi(x)$ we find

$$f(v,x) = \exp(-[mv^2 + 2\phi_0 e^{-x^2/w^2}]/2T) = \exp(-\tfrac{mv^2}{2T})\exp(-\tfrac{\phi_0}{T}e^{-x^2/w^2}). \quad (8.27)$$

If $\phi_0 > 0$, then no matter how small v^2 is, there is a real solution for v_∞ to the conservation equation

$$\frac{1}{2}mv^2 + \phi = \frac{1}{2}mv_\infty^2.$$

Therefore this expression for f is valid for all v. The distribution is everywhere Maxwellian, but its density varies with position. However, if $\phi_0 < 0$ then, everywhere that ϕ is negative, there is a minimum speed $\sqrt{-2\phi/m}$ below which there is no real solution for v_∞. These are the trapped orbits. They do not extend to infinity, but are reflected because they reside in the potential well. The value of f on those trapped orbits is undefined by the boundary condition at infinity, and must be determined otherwise, e.g. from the initial conditions. Fig. 8.8 shows an example solution.

Exercise 8. Boltzmann's equation

1. Divergence of acceleration in phase-space.
   (a) Prove that particles of charge q moving in a magnetic field $\boldsymbol{B}$ and hence subject to a force $q\boldsymbol{v} \times \boldsymbol{B}$, nevertheless have $\nabla_v.\boldsymbol{a} = 0$.
   (b) Consider a frictional force that slows particles down in accordance with $\boldsymbol{a} = -K\boldsymbol{v}$, where K is a constant. What is the "velocity-divergence" of this acceleration, $\nabla_v.\boldsymbol{a}$? Does this cause the distribution function f to increase or decrease as a function of time?
2. Write down the Boltzmann equation governing the distribution function of neutral particles of mass m in a gravitational field $g\hat{\boldsymbol{x}}$, moving through matter that consists of two different species of density n_a, n_b whose

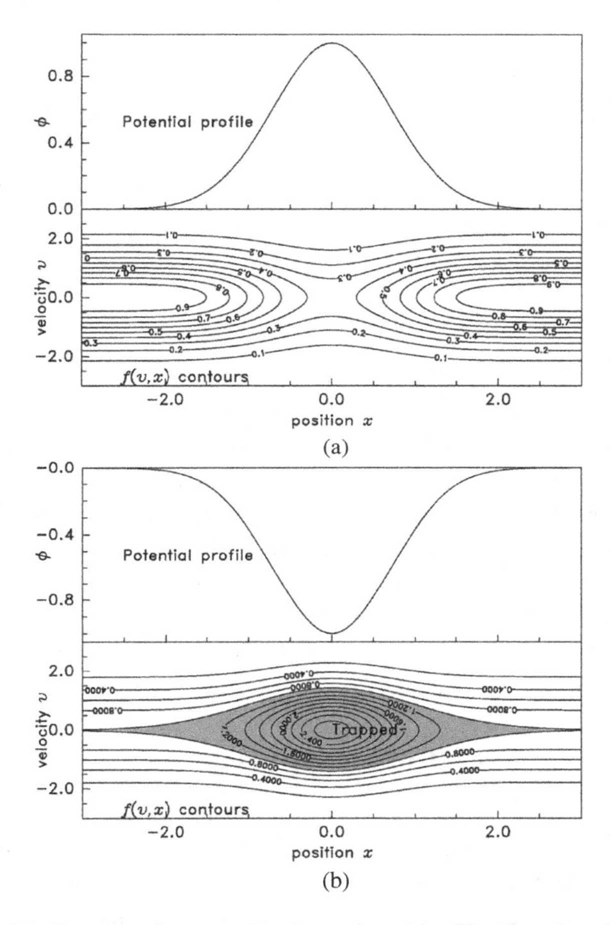

Figure 8.8 Contours of constant $f(v,x)$ are also orbits. Therefore the orbits can be plotted simply by contouring f, whose value is determined by the total (kinetic plus potential) energy at any point in phase-space. When the potential has a hill (a), all orbits extend to $x \to \infty$ and f is determined by boundary conditions. When the potential has a well (b), the value of f on trapped orbits (shaded) is undetermined. [The parameters used in these plots are $\phi_0/m = \pm 1$, $w = 1$, $T/m = 1$.]

only effects are: species a absorbs the particles with a cross-section σ_a independent of speed; species b emits the particles with a Maxwellian distribution of small temperature T_b, by radioactive decay with a half-life t_b.

Solve the equation (analytically) in uniform steady state ($\partial/\partial t = \partial/\partial \boldsymbol{x} = 0$), in the limit $kT_b \ll mg/n_a\sigma_a$, to find the distribution function $f_x(v_x)$.

9
Energy-resolved diffusive transport

When there are strong processes that drive the velocity distribution function away from equilibrium, it is generally important to account for the full distribution of velocities of particles to understand their transport. Sources of particles with kinetic energy substantially higher than the typical (e.g. thermal) energy will have this effect. Examples include all sorts of *reactions*; for example the chemical reactions that occur in combustion, or, as we will address in this chapter, the nuclear reactions that involve neutrons.[1]

9.1 Collisions of neutrons

Neutrons experience no net electric or magnetic forces, because they are uncharged, and usually gravity is ignorable; so in Boltzmann's equation the term proportional to the acceleration $\boldsymbol{a}$ can usually be neglected. Self-collisions are also negligible. It is the background matter through which the neutrons are moving that provides the targets with which the neutrons collide. It can be taken to consist of practically stationary atoms.

Collisions give the crucial terms in Boltzmann's equation. They arise from a whole host of different nuclear species, and the relevant cross-sections have very strong dependence on neutron kinetic energy (or equivalently speed, v). We generally sum over all the relevant species to give appropriate total source and sink rates per unit phase-space volume. In addition to the sink of neutrons, $-\Sigma_t vf$ from all possible collisions, there are sources arising from scattering,

[1] The transport of radiation, of *photons*, can also be treated in this way, but their treatment has to be expressed in terms of energy (or momentum) rather than speed, since all photons travel at the same speed.

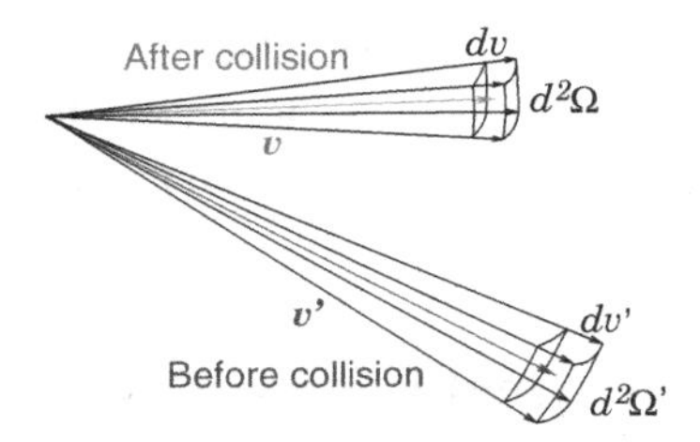

Figure 9.1 Collisions give a source of particles in $dvd^2\Omega$ as a result of particles that (before the collision) are in $dv'd^2\Omega'$. Both induced fissions (weighted by the average neutron yield ν) and scattering events contribute. The sink term in $dvd^2\Omega$ is the sum of all collisions that remove a particle from $dvd^2\Omega$.

and from fission reactions.[2] For fission and scattering, as illustrated by Fig. 9.1, we consider an incoming neutron denoted with a prime ($'$), having velocity given by speed v', and direction unit vector $\boldsymbol{\Omega}'$ (so $\boldsymbol{v}' = v'\boldsymbol{\Omega}$) giving rise to an outgoing (sourced) neutron with speed v, and direction $\boldsymbol{\Omega}$. The "macroscopic cross-sections" are written respectively Σ_f and Σ_s. They are functions of both incoming and outgoing velocity, which is denoted $(v' \to v, \boldsymbol{\Omega}' \to \boldsymbol{\Omega})$. Possible spatial dependence (on $\boldsymbol{x}$) is implicit. Each fission gives rise to an average number of outgoing neutrons ν typically greater than one, so for source the quantity required is actually $\nu\Sigma_f$. Also, to calculate the source we must integrate over all possible incoming velocities, for which we write the flux density[3] as $v'f'$ and the velocity element as $d^3v' = v'^2 d^2\Omega' \, dv'$. Thus Boltzmann's equation becomes the neutron transport equation:

$$\frac{\partial f}{\partial t} + v\boldsymbol{\Omega}.\frac{\partial f}{\partial \boldsymbol{x}} = \underbrace{-\Sigma_t v f}_{\text{sink}} + \int [\underbrace{\nu\Sigma_f(v' \to v, \boldsymbol{\Omega}' \to \boldsymbol{\Omega})}_{\text{fission}} + \underbrace{\Sigma_s(v' \to v, \boldsymbol{\Omega}' \to \boldsymbol{\Omega})}_{\text{scattering}}] v'f'v'^2 d^2\Omega' \, dv'. \tag{9.1}$$

This specific form of the collision term can also be used to treat collisions and chemical reactions between neutral molecules in a gas mixture.[4] For neutrons, the cross-sections are very complicated functions of speed v', and require

[2] Also there may be other spontaneous decays or fissions (delayed neutrons) but we'll ignore them for now.

[3] Most reactor physics texts express the velocity distribution differently, in terms of flux density $\phi \equiv vf$ as the dependent variable. To retain the more general relevance to solving Boltzmann's equation, we here keep f as the dependent variable. The result is still equivalent to standard reactor physics equations.

[4] Although there we might have to include self-collisions.

extensive data and careful integration to produce accurate collision terms, even when f is known. What's more, we have to deal with an integro-differential equation. It is not obvious how to solve it to find f self-consistently.

9.2 Reduction to multigroup diffusion equations

Solving six- or seven-dimensional integro-differential equations numerically is a major undertaking. If we just simple-mindedly discretize the distribution function $f(\boldsymbol{v}, \boldsymbol{x}, t)$ on finite grids in each dimension the amount of data quickly gets out of hand. Grids of length 100 require multiple terabytes of representation $100^6 = 10^{12}$, and solving for all of the discrete elements in phase-space becomes a grand computing challenge. Although there are some reasons to tackle that challenge, it is more usual, and historically more useful, instead to reduce the dimensionality of the problem by making appropriate choices of representation.

When the distribution function is nearly thermal, it is reasonable to describe it by just a few, low-order, moments of the velocity. This has the effect of reducing the three dimensions of velocity-space to just a few dependent parameters. They are the density $n = \int f d^3v$, the mean velocity $\int \boldsymbol{v} f d^3v/n$, and the mean kinetic energy per particle, equivalent to the temperature: $\int \frac{1}{2}mv^2 f d^3v/n = 3T/2$. Formally, taking the corresponding moments of Boltzmann's equation ends up giving us the standard fluid equations; continuity, momentum, and energy conservation. Thus, solving such a reduction has already been addressed in our discussion of the numerical treatment of fluid problems.

What do we do, though, when the velocity distribution is far from thermal, as it is in a reactor? We must keep account of that velocity dependence, because collision cross-sections depend upon it. A different type of approximation, useful in cases like neutron transport, because collisions are dominant, is to take the velocity *anisotropy* to be small. The distribution function f is nearly spherically symmetric: nearly independent of $\boldsymbol{\Omega}$; and it is unnecessary to represent the velocity-direction dependence of f in any detail.

It is necessary to retain just enough information about the anisotropy of $f(\boldsymbol{v})$ to represent the directed *flux density* of neutrons, which is what determines their transport. Consider neutrons of a specific speed v (i.e. in the element dv at v). Integrate the full transport equation (9.1) $v^2 d^2\Omega$ ($= d^3v/dv$ since $d^3v = v^2 d^2\Omega dv$) over the spherical velocity-space element, as illustrated in Fig. 9.2. We will denote that element just dv, to remind us of the speed choice, although its total velocity-space-volume is $4\pi v^2 dv$. Now we denote

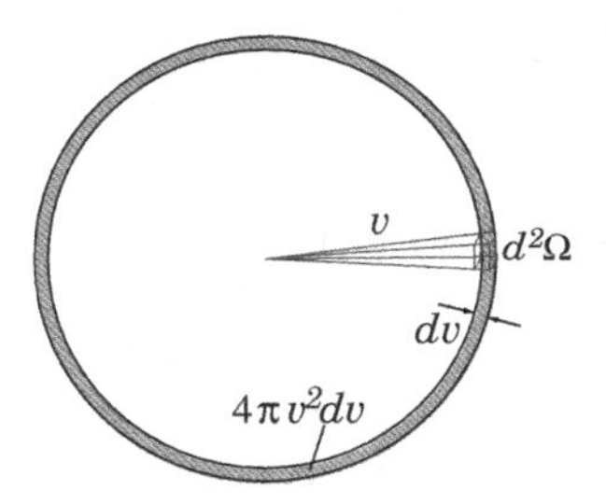

Figure 9.2 Illustrating the spherical volume in velocity-space over which the directional integral is performed.

the distribution function integrated over direction as

$$F(v) = \int_{dv} f v^2 d^2\Omega = \int_{dv} f d^3 v/dv. \tag{9.2}$$

So the number of particles per unit volume in the speed range dv is $F(v)dv$.

The first term in the angle-integrated Boltzmann equation becomes $\frac{\partial F}{\partial t}$. The second term becomes

$$\int_{dv} v\mathbf{\Omega}.\nabla f\, d^3v/dv = \nabla.[\int_{dv} \mathbf{\Omega} v f\, d^3v/dv] = \nabla.\mathbf{\Gamma}(v). \tag{9.3}$$

Here, $\mathbf{\Gamma}(v)dv = \int_{dv} \mathbf{\Omega} v f\, d^3v$ is the flux density of the particles in the speed element dv; so $\mathbf{\Gamma}(v)$ is the speed distribution of the flux density, a vector quantity.

Since all Σ are independent of $\mathbf{\Omega}$ (only the difference $\mathbf{\Omega} - \mathbf{\Omega}'$ matters), the first (sink) term on the right of eq. (9.1) becomes $-\Sigma_t vF(v)$ and the second (source term) can be written

$$Q(v) = \int [\nu \Sigma_f(v' \to v) + \Sigma_s(v' \to v)]v' F(v')dv', \tag{9.4}$$

in terms of Σ_f and Σ_s integrated over direction.

The direction-integrated transport equation is then

$$\frac{\partial F(v)}{\partial t} + \nabla.\mathbf{\Gamma}(v) = -\Sigma_t vF(v) + Q(v). \tag{9.5}$$

This equation becomes a diffusion equation if the flux density is proportional to the gradient of the density, an approximation usually called Fick's law. Written in terms of $F(v)$, this proportionality is

$$\mathbf{\Gamma}(v) = -D\nabla F(v), \tag{9.6}$$

where D is the "diffusion coefficient." The value of D is approximately $v/3\Sigma_t$, as outlined at the end of this section.

We then have the speed-resolved (or equivalently energy-resolved) diffusion equation

$$\frac{\partial F(v)}{\partial t} - \nabla.[D\nabla F(v)] = -\Sigma_t v F(v) + Q(v). \tag{9.7}$$

The lowest-order anisotropy of $f(\boldsymbol{v})$ is contained in $\boldsymbol{\Gamma} = D\nabla F(v)$, but the collision terms in this approximation are independent of any anisotropy. The equation applies for all values of the speed, v.

Enrichment: Derivation of diffusion coefficient.

The neutron (kinetic) transport eq. (9.1) can be turned into a spatial diffusion equation by considering a fixed speed v. To lowest order, we approximate the angular dependence of f as being a constant plus a term proportional to $\mu = \cos\theta$; i.e. as $f = f_0 + f_1\mu$, where f_0 and f_1 are independent of $\boldsymbol{\Omega}$. These are the first two terms of an expansion of the angular dependence in spherical harmonics. Reactor physics literature calls this the P_1 Approximation. Obviously this approximation only makes sense if the distribution has an approximate axis of (velocity) cylindrical symmetry, relative to which the polar angle θ is measured. In the absence of inherent material anisotropy, the local axis of symmetry must be in the direction of the density gradient ∇f_0.

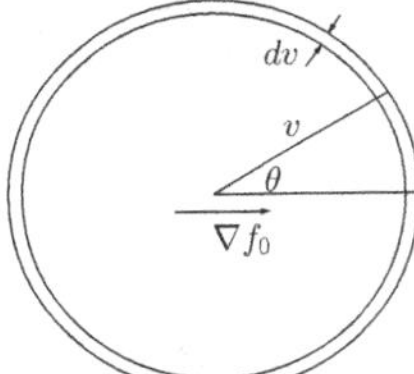

Integration over a spherical velocity element with $\mu = \cos\theta$ and the angle θ measured from the direction ∇f_0.

If we substitute into eq. (9.1) and equate orders of μ, ignoring the partial time derivative (since we presume the distribution in angle relaxes quickly), we then obtain at order μ

$$|\nabla f_0| = (-\Sigma_t + \Sigma_{s1})f_1,$$

where Σ_{s1} is the part of the scattering term proportional to μ, which is always smaller than Σ_t. (Sometimes Σ_{s1} is approximately ignorable.) Unlike scattering, fission does not contribute to this anisotropic component, except from its part in Σ_t, because it is generally presumed that there is no significant correlation between the direction of neutron emission from a fission event and the incoming neutron. [The degree of anisotropy in neutrons from fission reactions is low but formally non-zero, especially at higher incoming neutron energy.]

The contribution from the speed element dv to the directed spatial flux density (which is along the symmetry axis direction) is

$$\Gamma(v)dv = \int f v\mu d^3v = \int f v\mu\, 2\pi v^2 d\mu dv = 2\pi v^3 dv \int (f_0 + f_1\mu)\mu d\mu = \tfrac{1}{3} v f_1\, 4\pi v^2 dv.$$

Incorporating the previous equation, we then have

$$\boldsymbol{\Gamma}(v) = -\frac{v}{3(\Sigma_t - \Sigma_{s1})}\nabla(f_0 4\pi v^2).$$

This is Fick's law relating a flux density $\boldsymbol{\Gamma}dv$ to the gradient of a density $f4\pi v^2 dv$, times a diffusivity

$$D = \frac{v}{3(\Sigma_t - \Sigma_{s1})}.$$

9.3 Numerical representation of multigroup equations

We've made substantial progress in making the transport equation more managable, lowering its dimensionality from six phase-space dimensions to four (three space and one speed). Even so, we have to choose how to represent the distribution in speed (or energy), as well as the spatial representation.

9.3.1 Groups

The natural discretization in speed is to use ranges of speed or, equivalently, energy. In reactor physics the ranges are called "groups." It is like representing the speed distribution as a histogram (see Fig. 9.3). A point in phase-space (particle, if you like) belongs to the (integer) group g if its speed satisfies $v_{g-1/2} \leq v < v_{g+1/2}$. The half-integer-index speeds are the extrema of the speed range belonging to group g, and this group is regarded as having a typical or average speed v_g. Put another way, the group can be considered to be the integral over the finite speed element $\Delta v_g = v_{g+1/2} - v_{g-1/2}$. Take there to be N_G groups in all. Then each neutron group separately satisfies a diffusion equation like (9.7), except that the source integral $Q(v)$ for each group contains, in its integral, contributions from all the other groups, corresponding to fission neutrons appearing in one group when they were caused by another group, or scattering directly from one speed (group) to another:

$$\frac{\partial F_g}{\partial t} - \nabla.[D_g \nabla F_g] + \Sigma_{tg} v_g F_g = Q_g. \tag{9.8}$$

Since Q depends linearly on the F_g, through the integrals $\int v' F(v') dv'$, the discretized equations are naturally expressed in terms of a matrix equation

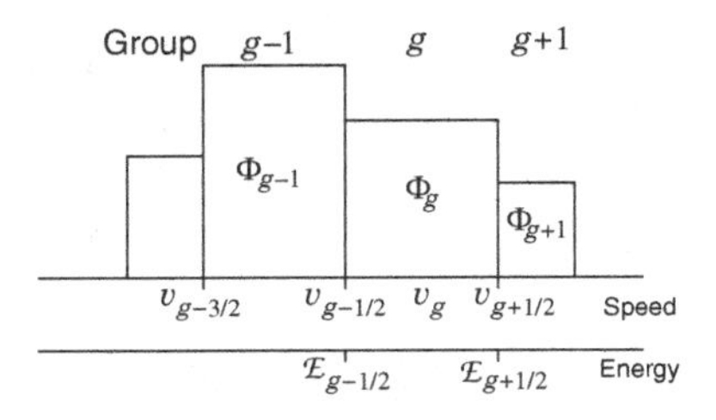

Figure 9.3 Groups are ranges in speed (equivalently energy) into which the neutrons are lumped. They are very rarely of equal width in speed or energy.

acting on a column vector of group fluxes

$$\frac{\partial \mathbf{F}}{\partial t} + (-\mathbf{L} + \mathbf{\Sigma}_t \mathbf{V})\mathbf{F} = \mathbf{Q} = \mathbf{A}\mathbf{F}. \tag{9.9}$$

The $N_G \times N_G$ matrices $\mathbf{L}$, $\mathbf{\Sigma}_t$ and $\mathbf{V}$ are diagonal, and their gth diagonal elements are $\nabla.D_g\nabla$, Σ_{tg}, and v_g, respectively. Matrix $\mathbf{A}$, which multiplies $\mathbf{F}$ to provide the source term $\mathbf{Q}$, is not diagonal. It couples together the different rows of the matrix equation, the different groups. The crucial point is that the collision matrices $\mathbf{\Sigma}_t$ and $\mathbf{A}$ can be calculated at any position $\boldsymbol{x}$ by performing appropriate integrals over speed v and then sums over different nuclear species. These require extensive information about the cross-sections for different types of collisions, but the velocity integrals only have to be done once.

Spatially uniform problem If we are treating an idealized "infinite homogeneous" reactor, then all the coefficients of the "leakage operator" $\mathbf{L}$ are zero: $\nabla.D_g\nabla = 0$. The spatial derivative terms are negligible. We need not represent more than one position in space, so the multiple group fluxes represent the only different dependent variable components we need to solve for. We have a first-order ordinary differential vector equation with time as the only independent variable.[5] In such a situation one might use a large number of speed groups N_G. The system can be solved by the methods of Chapter 2.

Non-uniform problem In an inhomogeneous or finite-sized reactor, the diffusive transport terms cannot be ignored. In principle, we can then discretize the reactor in space, producing a total of (say) N_S elements. Of course for a stuctured two-dimensional mesh, $N_S = N_1 \times N_2$, or for a three-dimensional; $N_S = N_1 \times N_2 \times N_3$. At each of the N_S spatial elements there are N_G groups, each of which has a speed-distribution component F_g. So there are a total of $N_S \times N_G$ $F_g(\boldsymbol{x})$-values to solve for.[6] In principle, we can line all these values up into a single column vector; then we can express the diffusion term ($\mathbf{LF}$) as finite differences between components adjacent in space. So it becomes a true multiplicative matrix rather than a matrix of differential operators.

We can in principle then advance the diffusion equation in time using an explicit scheme. In this situation the stability of our numerical scheme depends upon the diffusive nature of the D term. We need to recall the considerations

[5] In practice delayed neutrons introduce a severe complication.

[6] Computational size reduction, retaining spatial dependence, might call for a very small number of speed groups. Perhaps even a single group $N_G = 1$. Then the neutron transport is reduced to a single diffusion equation.

for parabolic diffusion equations from Chapter 5. An explicit forward in time, but centred in space (FTCS) scheme, requires us to satisfy the stability condition $\Delta t \leq \Delta x^2/2N_d D$ where N_d is the number of space dimensions. Now $D \approx \frac{1}{3}v\ell_c$, where ℓ_c is the collision mean-free-path. Consequently, the stability condition can be considered to be

$$\Delta\ell \equiv v\Delta t \leq \frac{3}{2N_d}\frac{\Delta x}{\ell_c}\Delta x. \tag{9.10}$$

The distance a neutron travels during the timestep must be less than Δx times a factor $3\Delta x/2N_d\ell_c$, for stability. Superthermal neutrons are liable to be the most limiting of Δt, because their speed v is greater (higher energy), and their ℓ_c is longer (smaller collision cross-section).

9.3.2 Steady-state eigenvalue

The most significant aspect of the neutron diffusion equation, as applied to a fission reactor, is that it is a homogeneous equation,[7] meaning that every term in eq. (9.9) is proportional to **F**. That is because in a reactor essentially all the neutrons are generated by the fission reactions caused by the neutron flux itself. The steady-state solution of a homogeneous equation is identically zero unless the multiplying matrix happens to be singular; in other words, unless its determinant is zero. So the condition for there to be a non-trivial steady solution, representing a steadily operating reactor, to the multigroup diffusion equations is:

$$\det(\mathbf{L} - \mathbf{\Sigma}\mathbf{V} + \mathbf{A}) = 0. \tag{9.11}$$

Such a condition does not come about by luck. It must be carefully arranged by adjusting the reactivity of the reactor using control rods and so on. If this condition of "criticality" is not satisfied, then the solution is *not steady*, the power is either increasing or decreasing with time. The way the condition is generally represented in the mathematics is by the idealized supposition that one has a way to adjust the effective neutron yield of all fission reactions; in particular, that they can be multiplied by a reactivity factor $1/k$. Remember that $Q(v)$ arose from two terms: scattering and fission. Write them separately as $\mathbf{\Sigma}_s\mathbf{V}\mathbf{F}$ and $\boldsymbol{\nu}\mathbf{\Sigma}_f\mathbf{V}\mathbf{F}$, respectively, where the diagonal matrix $\boldsymbol{\nu}$ has coefficients ν_g, which represent the number of neutrons (per fission reaction) that

[7] And the boundary conditions are also homogeneous. Of course "homogeneous" does not here mean uniform in space; it means having no constant terms.

arise with speed in the range of group g. Introduce the multiplicative reactivity factor k so that

$$\mathbf{Q} = \mathbf{AF} = \mathbf{\Sigma}_s\mathbf{VF} + \frac{1}{k}\boldsymbol{\nu}\mathbf{\Sigma}_f\mathbf{VF}. \tag{9.12}$$

Then the steady diffusion equation becomes a (generalized) eigenvalue problem:

$$\left[(-\mathbf{L} + \mathbf{\Sigma}_t\mathbf{V} - \mathbf{\Sigma}_s\mathbf{V}) - \frac{1}{k}\boldsymbol{\nu}\mathbf{\Sigma}_f\mathbf{V}\right]\mathbf{F} = 0, \tag{9.13}$$

which can also be written in terms of neutron flux $\mathbf{\Phi} \equiv \mathbf{VF}$ as

$$\left[(-\mathbf{L}\mathbf{V}^{-1} + \mathbf{\Sigma}_t - \mathbf{\Sigma}_s) - \frac{1}{k}\boldsymbol{\nu}\mathbf{\Sigma}_f\right]\mathbf{\Phi} = 0. \tag{9.14}$$

In general there are *some* values of k for which the determinant of this matrix equation is zero. They are the eigenvalues.[8] Actually we want only the eigensolution with the largest value of k. That corresponds to the mode that would be fastest growing (or slowest decaying) in the original time-dependent equation (9.9). If this largest k is greater than 1, then we had to *reduce* the neutron production rate by this factor relative to the original diffusion equation to get steady state. In other words there were too many reactions originally before we introduced k. A reactor with eigenvalue k greater than 1 is supercritical; the neutron flux will increase as a function of time. By the same argument an eigenvalue less than 1 is subcritical; neutrons decrease as a function of time.

How do we find the eigenvalue? Well, one way is simply to use a library routine designed to find generalized eigenvalues,[9] and plug in the matrices. However, this is liable to be a very inefficient approach unless that routine can use the fact that all the matrices are sparse. Even the collision matrices $\mathbf{\Sigma}_s$ and $\mathbf{\Sigma}_f$ are very sparse. They are full matrices *locally*, in the sense that they couple all the different speed groups together. But they have no cross terms between different spatial locations. Consequently they are diagonal with respect to the spatial indices. Another way of saying the same thing is that, if we arrange the order of indices in the giant vector $\mathbf{F}$ such that all the groups at a particular position are adjacent to one another, then we can consider each matrix to be

[8] Strictly speaking, the inverse of the eigenvalues of the matrix $(\boldsymbol{\nu}\mathbf{\Sigma}_f\mathbf{V})^{-1}(-\mathbf{L} + \mathbf{\Sigma}_t\mathbf{V} - \mathbf{\Sigma}_s\mathbf{V})$.

[9] A generalized eigenvalue is the solution to $(\mathbf{A} - \lambda\mathbf{B})\mathbf{v} = 0$ where both $\mathbf{A}$ and $\mathbf{B}$ are general matrices.

an $N_S \times N_S$ matrix of blocks that are each $N_G \times N_G$ submatrices. The form is illustrated in eq. (9.15):

$\mathbf{\Sigma}_s =$ 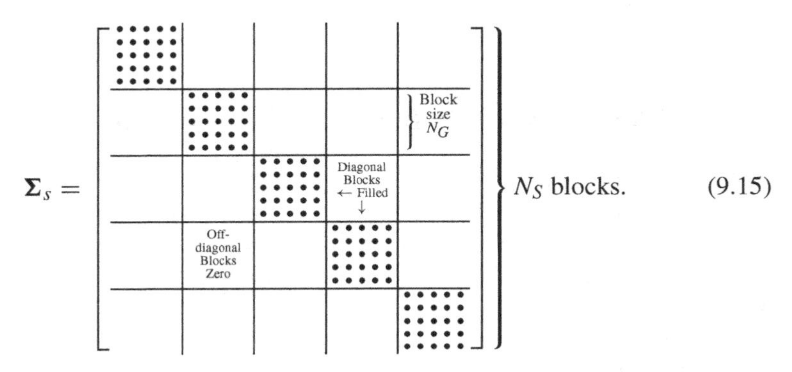

N_S blocks. (9.15)

In respect of the $N_S \times N_S$ structure, $\mathbf{\Sigma}_s$, $\mathbf{\Sigma}_f$, and $\mathbf{\Sigma}_t$ have diagonal arrangements of blocks. And $\mathbf{\Sigma}_t$ is diagonal even within its $N_G \times N_G$ blocks. By contrast, $\mathbf{L}$ is an arrangement of blocks that is tridiagonal, with additional subdiagonals in multiple space dimensions; like eq. (5.18) with each of the letter entries there representing a (diagonal) block.

Because of the sparseness, for a problem of any significant size, it is most efficient to use an *iterative* technique to find the eigenvalue, in which we need only to *multiply* by the original matrices, not to invert them. In this respect the problem is very similar to the challenge of solving large elliptic problems.

The fact that we are only interested in the largest eigenvalue is a big help.[10] Practically any iterative scheme that leads to the dominance of the corresponding eigenmode will work. Defining $-\mathbf{L} + \mathbf{\Sigma}_t\mathbf{V} - \mathbf{\Sigma}_s\mathbf{V} = \mathbf{M}$ and $\boldsymbol{\nu}\mathbf{\Sigma}_f\mathbf{V} = \mathbf{G}$ for brevity, a typical approach is to solve $[\mathbf{M} - \mathbf{G}/k]\mathbf{F} = 0$ using the scheme

$$\mathbf{M}\mathbf{F}^{(n+1)} = \frac{1}{k^{(n)}}\mathbf{G}\mathbf{F}^{(n)}. \tag{9.16}$$

At each outer step, n, an iterative solver[11] is used to find $\mathbf{F}^{(n+1)}$ given $\mathbf{F}^{(n)}$ (equivalent to inverting $\mathbf{M}$ but without actually forming $\mathbf{M}^{-1}$). Then the eigenvalue estimate is updated using a weighted ratio such as

$$k^{(n+1)} = \frac{(\mathbf{G}\mathbf{F}^{(n+1)})^T\mathbf{G}\mathbf{F}^{(n+1)}}{(\mathbf{G}\mathbf{F}^{(n+1)})^T\mathbf{M}\mathbf{F}^{(n+1)}}, \tag{9.17}$$

[10] Actually, if it were the largest eigenvalue of $\mathbf{M}$, for a single group or diagonal $\mathbf{G}$ we could directly use the "power method," which is eq. (9.16) with the F-indices $(n, n+1)$ exchanged. However, because we want the smallest eigenvalue $1/k$, we must effectively invert $\mathbf{M}$ (which requires interior iteration) because $\mathbf{M}$ is never diagonal.

[11] e.g. SOR.

and the step process is repeated[12]. As was discussed in respect of non-linear solvers, it might be advantageous to use only a very small number of loops of the inner iteration.

Worked example. Bare homogeneous reactor

Treat a reactor as having three neutron speed groups, whose material-interaction properties are uniform over the cuboid $0 < x < L_x$, $0 < y < L_y$, $0 < z < L_z$, with neutron density $F(v) = 0$ on the boundaries. The non-zero terms of the inverse collision length matrices (expressed in m^{-1}) may be taken as shown.[13]

Group (g) energy	1 10 keV–10 MeV	2 0.4 eV–10 keV	3 0–0.4 eV
$\Sigma_{tg} = v_g/3D_g$	20	53	94
$(\Sigma_t - \Sigma_s)_{gg}$	6.4	9.5	12
$(\Sigma_s)_{g+1,g}$	6.0	6.5	0
$(\nu\Sigma_f)_{1,g}$	0.9	1.8	18

So

$$\mathbf{DV}^{-1} = \begin{pmatrix} 0.015 & 0 & 0 \\ 0 & 0.0063 & 0 \\ 0 & 0 & 0.0035 \end{pmatrix} \mathrm{m}, \quad \mathbf{\Sigma}_t - \mathbf{\Sigma}_s = \begin{pmatrix} 6.4 & 0 & 0 \\ -6.0 & 9.5 & 0 \\ 0 & -6.5 & 12 \end{pmatrix} \mathrm{m}^{-1},$$

$$\text{and} \quad \boldsymbol{\nu}\mathbf{\Sigma}_f = \begin{pmatrix} 0.9 & 1.8 & 18 \\ 0 & 0 & 0 \\ 0 & 0 & 0 \end{pmatrix} \mathrm{m}^{-1}. \tag{9.18}$$

Find the reactivity eigenvalue and eigenmode for (a) a very large reactor, i.e. for $L_x, L_y, L_z \to \infty$, and (b) $L_x = L_y = L_z = 1$ m.

Although we could construct a large finite-difference block matrix and then numerically find its eigenvalue, spatially uniform material-interaction properties (collision lengths) constitute a very special case. They allow us to deduce the spatial variation of the eigenmode independently of its velocity

[12] See, e.g., A. Hébert (2009), *Applied Reactor Physics*, Presses Internationales Polytechnique, Montreal; see www.polymtl.ca. Actually many other choices of weighting will work as well, not just $(\mathbf{GF}^{(n+1)})^T$.

[13] The zeroes in the non-diagonal matrices arise because scatterings hardly ever move neutrons to higher energy or reduce the energy by more than one group, and fissions give rise only to energetic neutrons. The values given are roughly appropriate for a pressurized water reactor.

dependence. The velocity dependence and spatial dependence of the eigenmodes become *separable*, giving rise to a distribution function of the form

$$F(\boldsymbol{x}, v) = h(\boldsymbol{x})\Phi(v)/v, \tag{9.19}$$

where here h is independent of v and Φ is independent of space. This allows an enormous reduction in computational effort, because instead of one giant combined eigenvalue calculation, we need solve only two *separate*, much smaller, eigenvalue problems. The separate functions must satisfy the equation

$$\frac{1}{h}\nabla^2 h = B^2 = \frac{v}{D(v)\Phi(v)}[-\Sigma_t\Phi(v) + Q^{(k)}(v)], \tag{9.20}$$

where B^2 is the separation constant;[14] independent of both $\boldsymbol{x}$ and v. $Q^{(k)}$ denotes the source term modified by replacing the fission yield ν with ν/k.

For any specified reactor shape there is a set of eigenmodes that satisfy the boundary conditions and $\nabla^2 h = B^2 h$. This is a spatial eigenvalue problem for which B^2 is the eigenvalue. For a complicated shape of domain, it requires numerical solution, finding the eigenvalue of the finite-difference matrix form of the ∇^2 operator. In our simple cuboidal case, the spatial eigenmodes have the simple analytic form

$$h(\boldsymbol{x}) = \sin(\pi n_x x/L_x)\sin(\pi n_y y/L_y)\sin(\pi n_z z/L_z), \tag{9.21}$$

with n_x, n_y, n_z positive integers. For the mode with the longest wavelength ($n_x = n_y = n_z = 1$),

$$B^2 = \left(\frac{\pi}{L_x}\right)^2 + \left(\frac{\pi}{L_y}\right)^2 + \left(\frac{\pi}{L_z}\right)^2. \tag{9.22}$$

When we know B^2, the velocity-distribution eigenmode is then the solution of

$$\begin{aligned} 0 &= B^2(D(v)/v)\Phi(v) + \Sigma_t\Phi(v) - Q^{(k)}(v) \\ &= (B^2\mathbf{D}\mathbf{V}^{-1} + \boldsymbol{\Sigma}_t - \boldsymbol{\Sigma}_s - \tfrac{1}{k}\boldsymbol{\nu}\boldsymbol{\Sigma}_f)\boldsymbol{\Phi}, \end{aligned} \tag{9.23}$$

where the final form is the multigroup approximation expressed in terms of the size-N_G collision matrices.

For the large reactor, $B^2 \to 0$. Therefore, the multigroup eigenvalue problem is simply $[\boldsymbol{\Sigma}_t - \boldsymbol{\Sigma}_s]\boldsymbol{\Phi} = \frac{1}{k}[\boldsymbol{\nu}\boldsymbol{\Sigma}_f]\boldsymbol{\Phi} = 0$. Because of the special form of the matrices, one can quickly solve the equations by hand. They become $\Phi_2 = (12/6.5)\Phi_3$, $\Phi_1 = (9.5/6.0)\Phi_2$, and $6.4\Phi_1 - \frac{1}{k}[0.9\Phi_1 + 1.8\Phi_2 + 18\Phi_3] = 0$. Choosing to set $\Phi_1 = 1$, the eigenmode is then $\boldsymbol{\Phi}^T = (1, 0.632, 0.342)$, and the eigenvalue is $k_\infty = k = (0.9 + 1.8 \times 0.632 + 18 \times 0.342)/6.4 = 1.28$. This

[14] In reactor physics B is called the "Buckling."

k_∞ is the eigenvalue for an infinite-sized reactor. These values are confirmed if I enter the matrices into Octave and invoke the `eig` function. It tries to find three eigenmodes, but actually there is only one that is non-singular. It must be chosen correctly: a warning for the unwary.

For the 1-m reactor, $B^2 = 3\pi^2$, and we must add $B^2\mathbf{D}\mathbf{V}^{-1}$ to the matrix $\boldsymbol{\Sigma}_t - \boldsymbol{\Sigma}_s$. This does not change any of the zeroes, it just changes its diagonal entries somewhat, to $(6.84, 9.69, 12.1)$. Reevaluating the result we obtain eigenmode $\boldsymbol{\Phi}^T = (1, 0.619, 0.333)$, and eigenvalue $k = 1.20$. Thus, the finite domain size in this case introduces only very small changes in the energy spectrum of the reactor (the eigenmode), and reduces the eigenvalue only by a small amount. It is a general rule that increasing B^2 decreases k. Therefore, shorter-scale spatial modes (e.g. $n > 1$), for which B^2 is larger, always have less gain in a uniform reactor. To make this reactor operate at steady power, we would have to introduce control rods or make other adjustments to reduce the reactivity by the factor $1/k = 1/1.20$.

Exercise 9. Neutron transport

1. Consider a one-group representation of neutron transport in a slab, one-dimensional reactor of length $2L$. The reactor has uniform material properties; so that the steady diffusion equation becomes

$$-D\nabla^2 F + (\Sigma_t - S)F - \frac{1}{k}GF = 0,$$

where the diffusion coefficient (divided by velocity) D, the total attenuation "macroscopic cross-section" Σ_t, the scattering and fission source terms S, G, are simply scalar constants. F is the total neutron density, because there's only one group. For convenience, write $\Sigma_t - S = \Sigma$. The eigenvalue k must be found for this equation.

The boundary conditions at $x = \pm L$ are that the neutron density satisfy

$$F = -2D\hat{\mathbf{n}}.\nabla F = \mp 2D\frac{\partial F}{\partial x},$$

where $\hat{\mathbf{n}}$ is the outward normal at the boundary. This is essentially a non-reflective condition. It says there are no neutrons entering the reactor from outside.

Formulate the finite-difference diffusion equation on a uniform mesh of N_x nodes; node spacing $\Delta x = 2L/(N_x - 1)$. Exhibit it in the form of a matrix equation

$$\left[\mathbf{M} - \frac{1}{k}\mathbf{G}\right] \boldsymbol{F} = 0.$$

And write out the matrix $\mathbf{M}$ explicitly for the case $N_x = 5$ (so $\mathbf{M}$ is 5×5), carefully considering the incorporation of the finite-difference boundary condition.

The first and last rows of $\mathbf{M}$ that correspond to the boundary positions are not part of the eigensolution equation. In other words, the matrix $\mathbf{G}$ has zero first and last rows. Therefore, use the boundary conditions to *eliminate* F_1 and F_{N_x}, reducing the matrix dimension by two. Finally, arrive at a 3×3 eigenvalue equation $[\mathbf{M}' - G/k]\boldsymbol{F} = 0$, where $\mathbf{M}'$ is the 3×3 matrix adjusted to accommodate the boundary information, and G is just a scalar (equivalent to a factor times the unit matrix).

2. Implement this finite-difference scheme and (using some library function) find the eigenvalue, k, when $D = 1$, $\Sigma = 1$, $G = 1$, and $L = 2$ or $L = 10$. Use large enough N_x in your code that the solution is reasonably converged.

[Octave/Matlab hint. There are (in Octave) two routines for calculating eigenvalues: `eig()` and `eigs()`. Calling `eigs(M,K)` returns the largest `K` eigenvalues of the matrix `M`. Don't forget that the eigenvalue returned is λ solving $[\mathbf{M} - \lambda]\boldsymbol{\Phi} = 0$; in other words, it is the inverse of k. We want just the *smallest* λ, which corresponds to the largest k. We can trick this routine into giving it by using the generalized eigenvalue form $\mathbf{I} - k\mathbf{M} = 0$, so calling `eigs(eye(Nx),M,1)`. The `eigs()` routine uses an iterative technique. The `eig()` routine returns all the eigenvalues. It uses a direct solution technique. One then has to find the smallest, and invert it to give k.]

10

Atomistic and particle-in-cell simulation

When motivating the Boltzmann equation it was argued that there are too many particles for us to track them all, so we had to use a distribution-function approach. However, for some purposes, the number of particles that need to be tracked can in fact be managable. In that case, we can adopt a computational modeling approach that is called generically atomistic simulation. In short, it involves time-advancing individual classical particles obeying Newton's laws of motion (or in some relativistic cases Einstein's extensions of them).

There are also situations where it is advantageous to solve Boltzmann's equation by a method that uses pseudo-particles, each of them representative of a large number of individual particles. These pseudo-particles are modeled as if they were real particles. We'll come to this broader topic in a later section; for now, just think of real particles.

10.1 Atomistic simulation

If the particles are literally atoms or molecules, then since we will find it increasingly computationally difficult to track more than, say, a few billion of them, the volume of material that can be modeled is limited. A billion is $10^9 = 1000^3$; so we could at most model a three-dimensional crystal lattice roughly 1000 atoms on a side. For a solid, that means the size of the region we can model is perhaps 100 nm. Nanoscale. There's no way that we are going to atomistically model a macroscopic (say 1-mm) piece of material globally with forseeable computational capacity. Still, many very interesting and important phenomena relating to solid defects, atomic displacement due to energetic particle interactions, cracking, surface behavior, and so on, occur on the 100-nm scale. Materials behavior and design is a very important area for this direct atomic or molecular simulation.

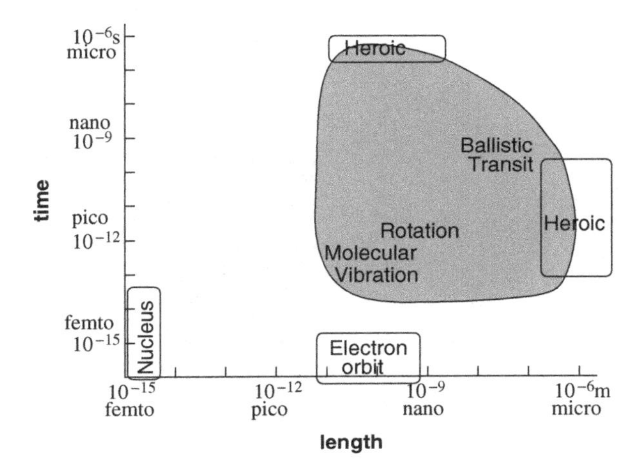

Figure 10.1 Approximate space- and time-scales for molecular dynamics atomistic simulation of condensed matter. Molecular vibration must be accommodated. Then computations of heroic effort are required to explore many orders of magnitude above it.

The time-scale in atomic interactions ranges from, say, the time it takes a gamma ray to cross a nucleus $\sim 10^{-23}$ s, to geological times of $\sim 10^{14}$ s. Modeling must choose a managable fraction of this enormous range, because the particle timesteps must be shorter than the fastest phenomenon to be modeled, and yet we can only afford to compute a moderate number of steps, maybe routinely as many as 10^4, but not usually 10^6, and only heroically 10^8. Phenomena outside our time-scale of choice have to be either irrelevant or represented by simplified representations of their effects in our modeling time-scale. Figure 10.1 illustrates the computationally feasible space- and time-scale region (shaded) indicating the approximate location of a few key phenomena.

For materials modeling where we are considering atoms whose thermal velocities are perhaps 1000 m/s, moving over lengths perhaps 10^{-7} m, the required rough time duration is 10^{-10} s for *molecular transit*. This is far longer than the characteristic time of motion of electrons within the atoms themselves, which is approximately the atomic size, 10^{-10} m, divided by the electron velocity at say 10 eV energy, 10^6 m/s: a time of 10^{-16} s for *electronic configuration*. This 1 million factor time range is too great to span routinely. So atomistic modeling usually needs to represent the atomic physics of the electron configurations in molecules in some averaged approximate way. This representation can sometimes be calculated on the basis of numerical solution of quantum mechanics, but we won't address that aspect of the problem. On the

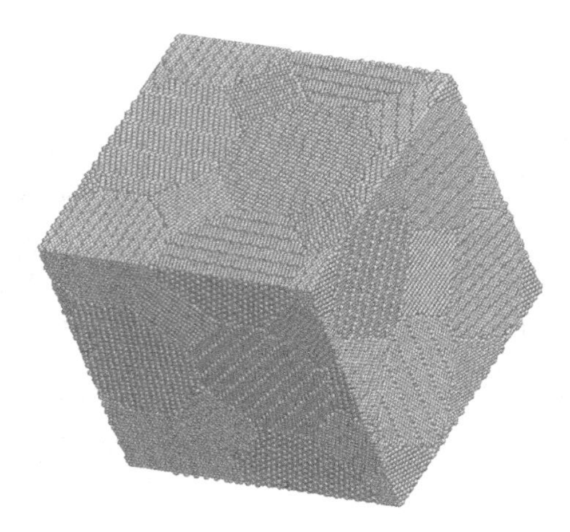

Figure 10.2 Example of a crystal lattice atomistic simulation in three dimensions. Study of a region of nanocrystalline metal with 840 000 atoms ready to be deformed. (Courtesy: Ju Li, Massachusetts Institute of Technology.)

other hand, the motions of the atomic nuclei arising from *molecular vibrations* have a typical time-scale of $\sim 10^{-13}$ s: at least 1000 times longer than electrons. This is manageable, and indeed *must* be resolved, if the dynamics of a lattice are to be modeled.

Atomistic modeling therefore represents the interatomic forces arising from electron orbital interactions *only approximately*, as time averages over electronic orbits or oscillations. But it follows the atoms themselves in their motions in response to the interatomic forces. The atoms are represented as classical particles interacting via a force field. This approach is sometimes called molecular dynamics. It dates from as early as 1956.[1] A modern example is shown in Fig. 10.2.

The way a simulation works then is, in outline:

1. Calculate the force at current position $\boldsymbol{x}$ on each particle due to all the others.
2. Accelerate and move particles for Δt, getting new velocities $\boldsymbol{v}$ and positions $\boldsymbol{x}$.
3. Repeat from 1.

Generally a fast second-order accurate scheme for the acceleration and motion (stage 2) is needed. One frequently used is the Leap-Frog scheme. Another is called the Verlet scheme, which can be expressed as

[1] B.S. Alder and T. E. Wainwright (1957), Phase transition for a hard sphere system, *J. Chem. Phys.* **27**, 1208-1209. Liquid behavior is an important phenomenon for which Molecular Dynamics is useful.

$$\begin{aligned} \boldsymbol{x}_{n+1} &= \boldsymbol{x}_n + \boldsymbol{v}_n \Delta t + \boldsymbol{a}_n \Delta t^2/2, \\ \boldsymbol{v}_{n+1} &= \boldsymbol{v}_n + (\boldsymbol{a}_n + \boldsymbol{a}_{n+1})\Delta t/2, \end{aligned} \tag{10.1}$$

where $\boldsymbol{a}_n$ is the acceleration corresponding to position $\boldsymbol{x}_n$.[2]

We also usually need to store a record of where the particles go, since that is the major result of our simulation. And we need methods to analyse and visualize the large amount of data that will result: the number of steps N_t times the number of particles N_p times at least six (three space, and three velocity) components.

10.1.1 Atomic/molecular forces and potentials

The simplest type of forces, but still useful for a wide range of physical situations, are particle-pair attraction and repulsion. Such forces act along the vector $\boldsymbol{r} = \boldsymbol{x}_1 - \boldsymbol{x}_2$ between the particle positions, and have a magnitude that depends only upon the distance $r = |\boldsymbol{x}_1 - \boldsymbol{x}_2|$ between them. An example might be the inverse-square electric force between two charges q_1 and q_2, $\boldsymbol{F} = (q_1 q_2/4\pi\epsilon_0)\boldsymbol{r}/r^3$. But for atomistic simulation more usually neutral particles are being modeled whose force changes from mutual attraction at longer distances to mutual repulsion at short distances. A very common form of interatomic potential that gives this kind of attraction and repulsion is the Lennard–Jones (12:6) potential (see Fig. 10.3(a))

$$\boldsymbol{F} = -\nabla U \qquad \text{with} \qquad U = \mathcal{E}_0 \left[\left(\frac{r_0}{r}\right)^{12} - 2\left(\frac{r_0}{r}\right)^{6} \right]. \tag{10.2}$$

A simplicity, but also a weakness, in the Lennard–Jones form is that it depends upon just two parameters, the typical energy $\mathcal{E}_0$, and the typical distance r_0. The equilibrium distance, corresponding to where the force $(-dU/dr)$ is zero, is r_0. At this spacing the binding energy is $\mathcal{E}_0$. The maximum attractive force occurs where $d^2U/dr^2 = 0$, which is $r = 1.109 r_0$ and it has magnitude $2.69\mathcal{E}_0/r_0$. The weakness of having only two parameters is that the spring-constant for the force, d^2U/dr^2, near the the equilibrium position cannot be set independent of the binding energy. An alternative force expression that allows this independence, by having three parameters, is the Morse form (see Fig. 10.3(b))

[2] For fixed Δt, the Verlet and Leap-Frog schemes are equivalent, with the identification $\boldsymbol{v}_n = (\boldsymbol{v}_{n-1/2} + \boldsymbol{v}_{n+1/2})/2$. The Verlet scheme, implemented in terms of velocity as in eq. (10.1), requires more storage or more acceleration evaluations: because two values of $\boldsymbol{a}$ are needed. It can be implemented in terms of the position advance alone as $\boldsymbol{x}_{n+1} = 2\boldsymbol{x}_n - \boldsymbol{x}_{n-1} + \boldsymbol{a}\Delta t^2$, which requires the same storage as the Leap-Frog scheme.

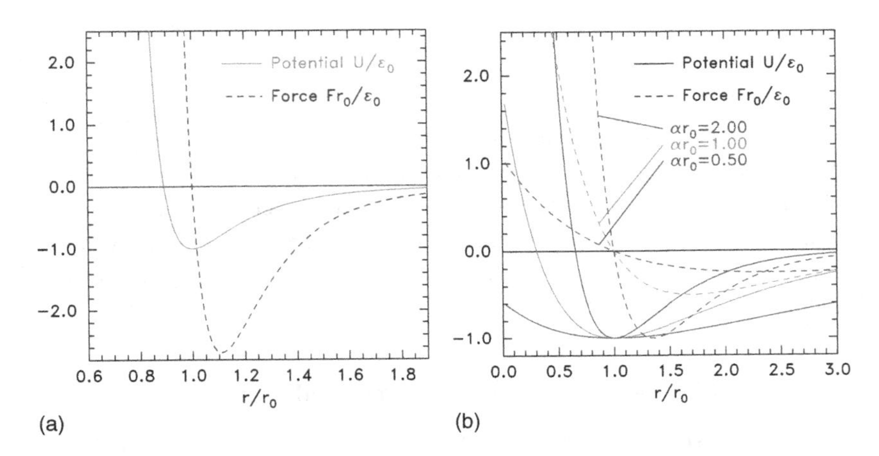

Figure 10.3 Potential and corresponding force forms for (a) Lennard–Jones (eq. (10.2)) and (b) Morse (eq. (10.3)) expressions.

$$U = \mathcal{E}_0 \left(e^{-2\alpha(r-r_0)} - 2e^{-\alpha(r-r_0)} \right). \tag{10.3}$$

Its force is zero at r_0, where the binding energy is $\mathcal{E}_0$; but the spring-constant there can be adjusted using the parameter α. See Fig. 10.3(b).

These simple two-particle, radial-force, forms omit several phenomena that are important in molecular interactions in nature. These additional phenomena include higher-order potential interactions, represented by the total potential energy of the entire assembly being given as a heirarchy of sums of multiple-particle interactions

$$U = \sum_i U_1(\boldsymbol{x}_i) + \sum_{ij} U_2(\boldsymbol{x}_i, \boldsymbol{x}_j) + \sum_{ijk} U_3(\boldsymbol{x}_i, \boldsymbol{x}_j, \boldsymbol{x}_k) + \dots, \tag{10.4}$$

where the subscripts $i, j, \dots$ refer to different particles. The force on a particular particle l is then $-\frac{\partial}{\partial \boldsymbol{x}_l} U$. The first term U_1 represents a background force field. The second represents the pairwise force that we've so far been discussing. We've been considering the particular case where U_2 depends only on $r = |\boldsymbol{x}_i - \boldsymbol{x}_j|$. The third (and higher) terms represent possible multi-particle correlation forces. They are often called "cluster" potential terms.

Other force laws between multi-atom molecules might include the orientation of the molecule bonds. In that case, internal orientation parameters would have to be introduced or else the molecule's individual atoms themselves represented by particles whose bonds are modeled using force laws appropriate to them. They could be represented as a third- or probably at least fourth-order series for the potential.

10.1.2 Computational requirements

If there are N_p particles, then to evaluate the force on particle i from all the other particles j requires N_p force evaluations for pair-forces (U_2 term). It requires N_p^2 for three-particle (U_3) terms, and so on. The force calculation needs to be done for all the particles at each step, so if we include even just the pair-forces for all particles, N_p^2 force terms must be evaluated. This is too much. For example, a million particles would require 10^{12} pair-force evaluations per timestep. Computational resources would be overwhelmed. Therefore, the most important enabling simplification of a practical atomistic simulation is to reduce the number of force evaluations till it is not much worse than linear in N_p. This can be done by taking advantage of the fact that the force laws between neutral atoms have a rather short range; so the forces can be ignored for particle spacings greater than some modest length. In reality, we only need to calculate the force contributions from a moderately small number of nearby particles on each particle i. It is not sufficient to look at the position of all the other particles at each step and decide whether they are near enough to worry about. That decision is itself an order N_p cost per particle (N_p^2 total). Even if it's a bit cheaper than actually evaluating the force, it won't do. Instead, we have to keep track, in some possibly approximate way, of which of the other particles are close enough to the particle i to matter to it.

There are broadly two ways of doing this. Either we literally keep a list of near neighbors associated with each particle. Or else we divide up the volume under consideration into much smaller blocks and adopt the policy of only examining the particles in its own block and the neighboring blocks. Either of these will obviously work for a crystal-lattice-type problem, modeling a solid, because the atoms hardly ever change their nearest neighbors, or the members of blocks. But in liquids or gases the particles can move sufficiently that their neighbors or blocks are changing. To recalculate which particles are neighbors costs $\sim N_p^2$. However, there are ways to avoid having to do the neighbor determination every step. If we do it rarely enough, we reduce the cost scaling.

Neighbor list algorithm A common way to maintain an adequately accurate neighbor list is as follows. See Fig. 10.4. Suppose r_c is the cut-off radius beyond which interparticle forces are negligible. For each particle i, designate as neighbors particles within a larger sphere $|\boldsymbol{x}_j - \boldsymbol{x}_i| = r < r_l$. Suppose the fastest speed a particle has is v_{max}; then we know that no particle, starting from outside the sphere r_l, can reach the sphere r_c in less than a time $(r_l - r_c)/v_{max}$, that is, in fewer timesteps than $N_l = (r_l - r_c)/v_{max}\Delta t$. Consequently, we need to

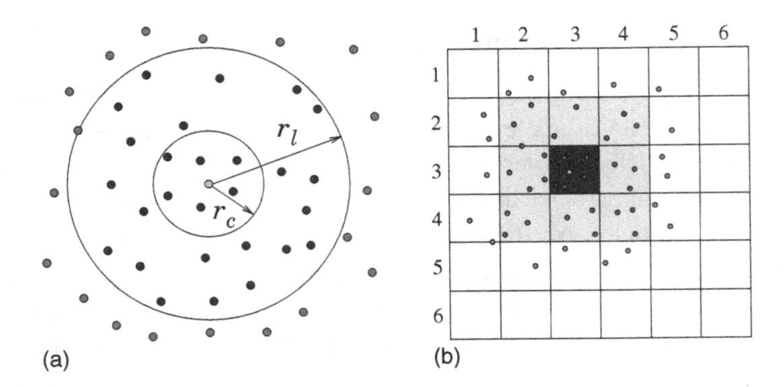

Figure 10.4 Neighbor list (a), and block (b), algorithms enable testing only nearby particles for force influence. The neighbors whose influence must be calculated are inside the radius r_l or inside the shaded region.

update the neighbor lists only every N_l steps. The average neighbor list cost per step is N_p^2/N_l. This is smaller than N_p^2, and if N_l is large (because the maximum velocity is small), much smaller than N_p^2. However, it is still of order N_p^2.[3]

Block algorithm If the domain is divided into a large number of blocks, each of which is bigger than the cut-off radius, then we need to examine only particles that are in the same block or an adjacent block. Adjacent blocks must be examined because a particle near the boundary of a block might be influenced by particles just the other side of that boundary. Suppose there are N_b blocks. They contain on average N_p/N_b particles each. As the size of the computational region increases, we can keep this ratio constant. The total number of neighboring particles we need to examine for each particle i is (in three dimensions) $3^3 N_p/N_b \propto const$. Thus, this block algorithm's step cost is $\propto N_p$, linear in the number of particles. But the constant of proportionality might be quite large. There is also an interesting question as to how the list of particles in a block is maintained. One way to do this is to use a linked list of pointers. However, such a linked list does not lend itself readily to parallel data implementations, and there are interesting forefront research questions as to the best practical way of solving this problem.

[3] Since the step cost for a neighboring domain of size $\propto N_l$ is $\propto N_p N_l^3$, and the step cost for the neighbor update is $\propto N_p^2/N_l$, a formal optimum occurs when these are equal; so that $N_l \propto N_p^{1/4}$. So formally, the per-step cost of this N_l-optimized algorithm would be $\propto N_p N_p^{3/4}$: slightly better than N_p^2.

10.2 Particle-in-cell codes

If the interparticle force law is of infinite range, as it is, for example, with the inverse-square interactions of charged particles in a plasma, or gravitating stars, then the near-neighbor reduction of the force calculation does not work, because there is no cut-off radius beyond which the interaction is negligible. This problem is solved in a different way, by representing the long-range interactions as the potential on a mesh of cells. This approach is called "particle in cell" or PIC.[4]

Consider, for simplicity, a single species of charged particles (call them electrons) of charge q ($= -e$ for electrons, of course) and mass m. Positive ions could also be modeled as particles, but for now take them to be represented by a smooth neutralizing background of positive charge density $-n_i q$. The electrons move in a region of space divided into cells labelled with index j at positions x_j. [Most modern PIC production codes are multidimensional, but the explanations are easier in one dimension.] They give rise to an electric potential ϕ. Ignoring the discreteness of the individual electrons, there is a smoothed-out total (sum of electron and ion) charge density $\rho_q(x) = q[n(x) - n_i]$. The potential satisfies Poisson's equation,

$$\nabla^2\phi = \frac{d^2\phi}{dx^2} = -\frac{\rho_q}{\epsilon_0} = -\frac{q[n(x) - n_i]}{\epsilon_0}. \tag{10.5}$$

We represent this potential discretely on the mesh: ϕ_i and solve it numerically using a standard elliptic solver. The only new feature is that we need to obtain a measure of smoothed-out density on the mesh. We do this by assigning the charge density of the individual electrons to the mesh in a systematic way. The simplest way to assign it is to say that each electron's charge is assigned to the *nearest grid point* NGP. That's equivalent to saying each electron is like a rod of length equal to the distance Δx between the grid points, and it contributes charge density equal to $q/\Delta x$ from all positions along its length. See Fig. 10.5. The volume of the cell is Δx and the electron density is equal to the number of particles whose charge is assigned to that cell divided by the cell volume. Usually a continuous linear interpolation is preferred, called the *cloud in cell* (CIC) assignment. The charge density assigned from each electron is equal to $q/\Delta x$ when the electron is exactly at x_j, and falls linearly, reaching zero when the particle is at $x_{j\pm1}$. Thus, the electron is like a rod of length $2\Delta x$ whose charge distribution is triangular.

[4] See for example C. K. Birdsall and A. B. Langdon (1991), *Plasma Physics via Computer Simulation*, IOP Publishing, Bristol; or R. W. Hockney and J. W. Eastwood (1988), *Computer Simulation using Particles*, Taylor and Francis, New York.

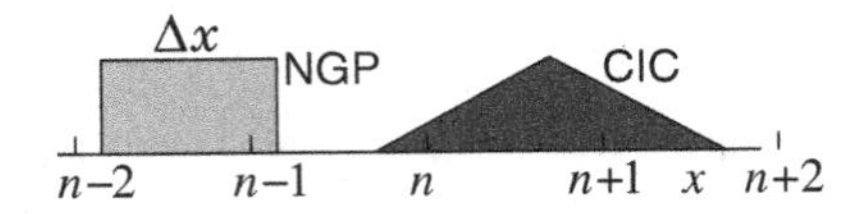

Figure 10.5 Effective shapes for the NGP and CIC charge assignments.

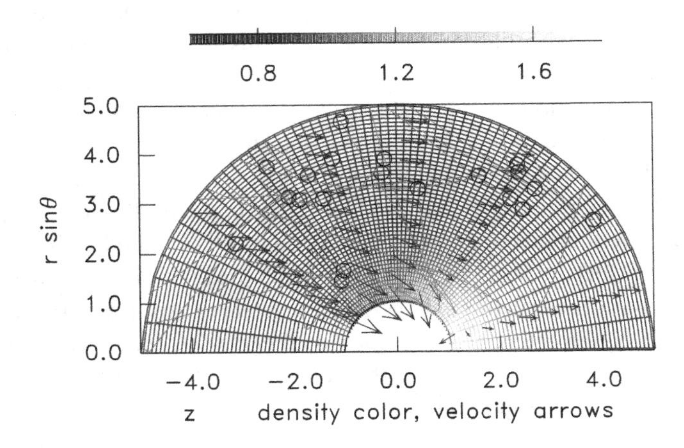

Figure 10.6 A curved grid (relatively unusual for PIC) shaded to represent the density normalized to the distant value. A few representative particle orbits in the vicinity of a spherical object are shown, and arrows indicate the mean ion velocity.

The way the PIC code runs is this.

1. Assign the charge from all the particles onto the grid cells.
2. Solve Poisson's equation to find the potential ϕ_j.
3. For each particle, find $\nabla\phi$ at its position x_i by interpolation from the x_j.
4. Accelerate, by the corresponding force, and move the particles.
5. Repeat from 1.

This process will simulate the behavior of the plasma accounting realistically for the particle motion. So it is a kind of atomistic simulation. Fig. 10.6 shows an illustration of an example in a spherical computational domain.

Why introduce this cell mesh? Because this approach is computationally far more efficient than adding up the inverse-square law forces between individual particles. An atomistic pair-force approach costs $\sim N_p^2$ per step. By contrast, the particle moving stage, once the force is known, is $\sim N_p$. If the potential grid has a total of N_g points, then an efficient iterative Poisson solution costs $\sim N_g N_g^{1/N_d}$ per step, in N_d dimensions,[5] or can be done by tridiagonal elimination in $\sim N_g$ operations in one dimension. Generally, the number of particles per cell is fairly large, so N_g is much smaller than N_p and the Poisson cost scales linearly or nearly linearly with N_g. Therefore, for practical purposes, the costs are dominantly those of interpolating the electric field to the particle and moving it: an order N_p cost, not N_p^2 like the pair-force approach.

Sometimes the dynamics of the ions is just as important to model as the electrons. Then the ions must be treated through the PIC approach as a second species of particles obeying Newton's law. Actually it is sometimes advantageous to treat only the ions this way, and treat electrons as a continuum whose density is a known function of ϕ. The latter approach is often called "hybrid" PIC.

10.2.1 Boltzmann equation pseudo-particle representation

In a PIC code, the particles move and are tracked in phase-space: $(\boldsymbol{x}, \boldsymbol{v})$ is known at each timestep. A particle's equation of motion in phase-space is

$$\frac{d}{dt}\begin{pmatrix}\boldsymbol{x}\\ \boldsymbol{v}\end{pmatrix} = \begin{pmatrix}\boldsymbol{v}\\ \boldsymbol{a}\end{pmatrix}. \tag{10.6}$$

This is also the equation of motion of the *characteristics* of the Boltzmann equation (8.12, 8.13). Thus, advancing a PIC code with N_p particles is equivalent to integrating along N_p characteristics of the Boltzmann equation. But what about collisions?

The remarkable thing about a PIC code, in its simplest implementation, is that it has essentially *removed* all charged-particle collisions. The grainy discreteness of the electrons is smoothed away by the process of assigning charge to the grid and then solving for ϕ. Therefore, unless we do something to put collisions back, the PIC code actually represents integration along characteristics of the *Vlasov* equation, the *collisionless* Boltzmann equation. If we had instead used (highly inefficiently) pair-forces, then the charged-particle collisions would have been retained.

Because the collisions have been removed from the problem, the actual magnitude of each particle's charge and mass no longer matters; only the *ratio* q/m appears in the acceleration $\boldsymbol{a}$ in the Vlasov equation. That means we can represent physical situations that would in nature involve unmanagable numbers of physical electrons by regarding the electrons (or ions) of our computation as *pseudo-particles*. Each pseudo-particle corresponds to a

[5] See for example the successive over-relaxation (SOR) estimates.

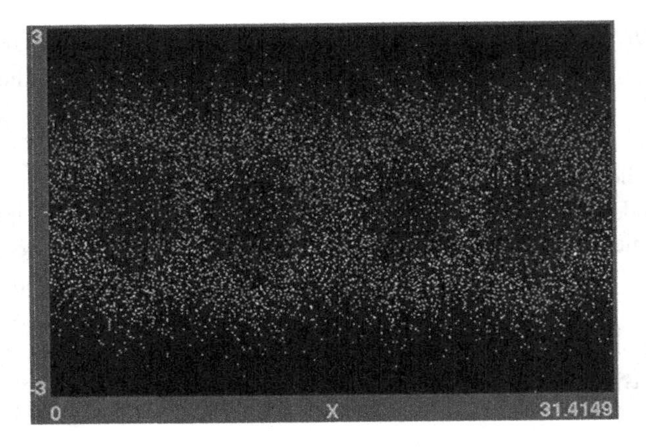

Figure 10.7 Example of phase-space locations of electrons. A one-dimensional v versus x calculation is illustrated using the code XES1 (by Birdsall, Langdon, Verboncoeur and Vahedi, originally distributed with the text by Birdsall and Langdon) for two streams of particles giving rise to an instability whose wavelength is four times the domain length. Each electron position is marked as a point. Their motion can be viewed like a movie.

very large number of actual particles, reducing the number of computation pseudo-particles to a manageable total, and keeping the costs of the computation within tolerable bounds. For our computation to remain a faithful representation of the physical situation, we require only that the resolution in phase-space, which depends upon the total number of randomly distributed electrons, should be sufficient for whatever phenomenon is being studied. We also require, of course, that the potential mesh has sufficient spatial resolution.

PIC codes are a backbone of much computational plasma physics, important for modeling semiconductor processing tools, space interactions, accelerators, and fusion experiments. An example of a one-dimensional PIC calculation is shown in Fig. 10.7. They are particularly useful for collisionless or nearly collisionless problems that are widespread in the field. They can also be modified to include collisions of various different types, as the conditions require them. In plasmas, though, charged-particle collisions are often dominated by small scattering angles and are much better approximated by a Fokker–Planck diffusion in phase-space than by discrete events.

10.2.2 Direct-simulation Monte Carlo treatment of gas

An approach that combines some of the features of PIC and atomistic simulation is the treatment of tenuous neutral gas behavior by what have come to be called direct-simulation Monte Carlo (DSMC) codes. These are for addressing situations where the ratio of the mean-free-path of molecules to the characteristic spatial feature size (the "Knudsen number") is of order unity (within a factor of 100 or so either way). Such situations occur in very tenuous gases (e.g. orbital re-entry in space) or when the features are microscopic. DSMC shares with PIC the features that the domain is divided into a large number of cells, that pseudo-particles are used, and that collisions are represented in a simplified way that reduces computational cost and yet approximates physical behavior. DMSC is also, in effect, integrating the Boltzmann equation along characteristics, but in this case there's no acceleration term, so the characteristics are straight lines.

The pseudo-particles representing molecules are advanced in time, but at each step, chosen to be somewhat shorter than a typical collision time, they are examined to decide whether they have collided. In order to avoid a N_p^2 cost, collisions are considered only with the particles in the same cell of the grid. (This partitioning is all the cells are used for.) The cells are chosen to have a size smaller than a mean-free-path, but not by much. They will generally have only a modest number (perhaps 20–40) of pseudo-particles in each cell. The number of individual molecules represented by each pseudo-particle is adjusted to achieve this number per cell.

Whether a collision has occurred between two particles is decided based only upon their relative *velocity*, not on their position within the cell. This is the big approximation. A statistical test using random numbers decides if and which collisions have happened. A collision changes the velocity of both colliding particles, in accordance with the statistics of the collision cross-section and corresponding kinematics. That way, momentum and energy are appropriately conserved within the cell as a whole. Steps are iterated, and the overall behavior of the assembly of particles is monitored and analysed to provide effective fluid parameters like density, velocity, effective viscosity, and so on. Fig. 10.8 shows an example from the code DSMC, v3.0 developed by Graeme Bird.

10.2.3 Particle boundary conditions

Objects that are embedded in a particle computation region present physical boundaries at which appropriate approximate conditions must be applied. For example with DSMC, gas particles are usually reflected, whereas with plasmas it is usually assumed that the electrons are removed by neutralization when they encounter a solid surface.

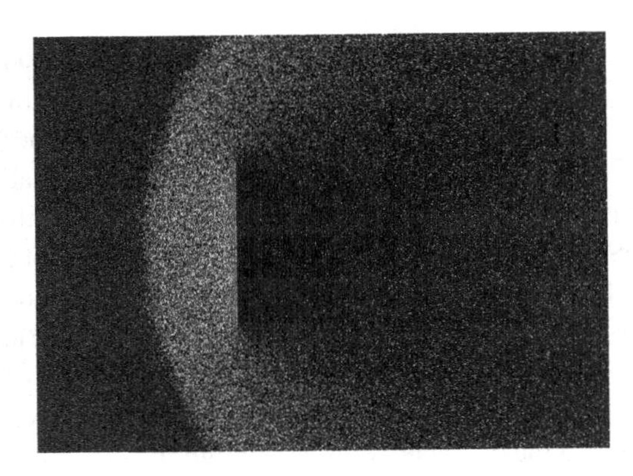

Figure 10.8 Example of position plot in two space dimensions of tenuous gas flow past a plate. Different colors (shadings) indicate molecules that have been influenced by the plate through collisions.

An important question arises in most particle simulation methods. What do we do at the outer boundary of our computational domain? If a particle leaves the domain, what happens to it? And what do we do to represent particles entering the domain?

Occasionally the boundary of our domain might be a physical boundary no different from an embedded object. But far more often the edge of the domain is simply the place where our computation stops, not where there is any specific physical change. What do we do then?

The appropriate answer depends upon the specifics of the situation, but quite often it makes sense to use *periodic boundary conditions*. Periodic conditions for particles are like periodic conditions for differential equations, discussed in Section 3.3.2. They treat the particles as if a boundary that they cross corresponds to the same position in space as the opposite boundary. A particle moving on a computational domain in x that extends from 0 to L, when it steps past L, to a new place that would have been $x = L + \delta$, outside the domain, is reinserted at the position $x = \delta$, close to the opposite boundary, but back inside the domain. Of course the particle's velocity is just what it would have been anyway. Velocity is not affected by the reinsertion process. Periodic conditions can be applied in any number of dimensions.

Periodic boundaries mean that the computation represents the phenomena of a periodic array of domains all connected to one another and all doing the same thing. Sometimes that is actually what one wants. But more often it is an approximation to a larger domain. If nothing of interest happens at a scale equal to or larger than the smaller computational domain, then the artificially imposed periodicity is unimportant, and the periodic conditions are

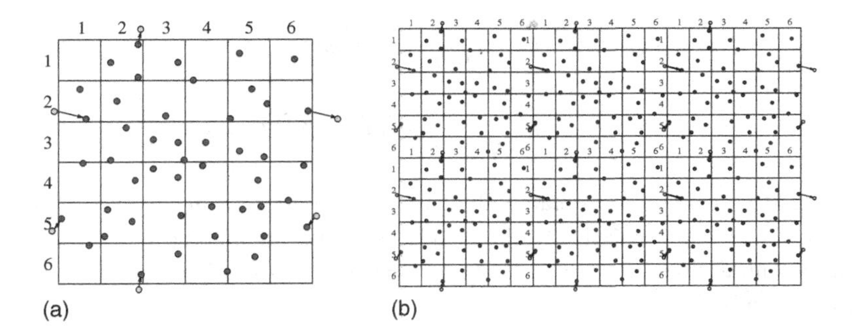

Figure 10.9 Particles that cross outer periodic boundaries (a) are relocated to the opposite side of the domain. This is equivalent (b) to modeling an (infinite) periodic array made up of repetitions of the smaller domain.

a convenient way to represent the computation of a small subvolume within a much larger, uniform, medium. Fig. 10.9 illustrates this point.

Sometimes, however, it is not appropriate to use periodic conditions. In that case a particle that leaves the domain is simply removed from the calculation. If the calculation is approximately steady, then clearly there must also be particles created within the domain or *entering* it from outside. A suitable method for injecting them into the calculation must be implemented. It might represent, for example, the flux of particles across the boundary from an assumed velocity distribution function.

Worked example. Required resolution of PIC grid

How fine must the potential mesh be in an electron PIC code?

Well, it depends on how fine-scale the potential variation might be. That depends on the parameters of the particles (electrons). Suppose they have approximately a Maxwell–Boltzmann velocity distribution of temperature T_e. We can estimate the finest scale of potential variation as follows. We'll consider just a one-dimensional problem. Suppose there is at some position x a perturbed potential $\phi(x)$ such that $|\phi| \ll T_e/e$, measured from a chosen reference $\phi_\infty = 0$ in the equilibrium background where the density is $n_e = n_\infty = n_i$. (Referring to the background as ∞ is a helpful notation that avoids implying the value at $x = 0$; it means the distant value.) Then the electron density at x can be deduced from the fact that $f(v)$ is constant along orbits (characteristics). In the collisionless steady state, energy is conserved; so for any orbit $\frac{1}{2}mv^2 - e\phi = \frac{1}{2}mv_\infty^2$, where v_∞ is the velocity on that orbit when it is "at infinity" in the background, where $\phi = 0$. Consequently,

$$f(v) = f_\infty(v_\infty) = n_\infty \sqrt{\frac{m}{2\pi T_e}} \exp(-mv_\infty^2/2T_e)$$
$$= n_\infty \sqrt{\frac{m}{2\pi T_e}} \exp(-mv^2/2T_e + e\phi/T_e).$$

Hence at $x, f(v)$ is Maxwellian with density $n = \int f(v)dv = n_\infty \exp(e\phi/T_e)$.

Now let's find analytically the steady potential arising for $x > 0$ when the potential slope at $x = 0$ is $d\phi/dx = -E_0$. Poisson's equation in one dimension is

$$\frac{d^2\phi}{dx^2} = -\frac{en_\infty}{\epsilon_0}\left[1 - \exp(e\phi/T_e)\right] \approx \left(\frac{e^2 n_\infty}{\epsilon_0 T_e}\right)\phi. \tag{10.7}$$

The final approximate form gives Helmholtz's equation. It is obtained by Taylor expansion of the exponential to first order, since its argument is small. The solution satisfying the condition at $x = 0$ is then

$$\phi(x) = E_0 \lambda e^{-x/\lambda}, \quad \text{where} \quad \lambda^2 = \left(\frac{e^2 n_\infty}{\epsilon_0 T_e}\right). \tag{10.8}$$

Based on this model calculation, the length $\lambda = \sqrt{e^2 n_\infty/\epsilon_0 T_e}$, which is called the Debye length, is the characteristic spatial scale of potential variation. A PIC calculation must have a fine enough grid to resolve the smaller of λ and the characteristic feature size of any object in the problem whose influence introduces potential structure. In short, $\Delta x \leq \lambda$.

If we had an operational PIC code, we could do a series of calculations with different cell size Δx. We would find that when Δx became small enough, the solutions would give a result independent of Δx. That would be a good way of demonstrating adequate spatial resolution numerically. For the simple problem we've considered the requirement can be calculated analytically. Actually the criterion $\Delta x \lesssim \lambda$ applies very widely in plasma PIC calculations.

Exercise 10. Atomistic simulation

1. The Verlet scheme for particle advance is

$$\begin{aligned} \boldsymbol{x}_{n+1} &= \boldsymbol{x}_n + \boldsymbol{v}_n \Delta t + \boldsymbol{a}_n \Delta t^2/2 \\ \boldsymbol{v}_{n+1} &= \boldsymbol{v}_n + (\boldsymbol{a}_n + \boldsymbol{a}_{n+1})\Delta t/2. \end{aligned} \tag{10.9}$$

Suppose that the velocity at integer timesteps is related to that at half-integer timesteps by $\boldsymbol{v}_n = (\boldsymbol{v}_{n-1/2} + \boldsymbol{v}_{n+1/2})/2$. With this identification, derive the

Verlet scheme from the Leap-Frog scheme,

$$\begin{aligned} \boldsymbol{x}_{n+1} &= \boldsymbol{x}_n + \boldsymbol{v}_{n+1/2}\Delta t \\ \boldsymbol{v}_{n+3/2} &= \boldsymbol{v}_{n+1/2} + \boldsymbol{a}_{n+1}\Delta t, \end{aligned} \tag{10.10}$$

and thus show that they are equivalent.

2. A block algorithm is applied to an atomistic simulation in a cubical three-dimensional region, containing $N_p = 1\,000\,000$ atoms approximately uniformly distributed. Only two-particle forces are to be considered. The cut-off range for particle–particle force is four times the average particle spacing. Find
   (a) The optimal size of blocks into which to divide the domain for fastest execution.
   (b) How many force evaluations per time-step will be required.
   (c) If the force evaluations require five multiplications, a Verlet advance is used, and the calculation is done on a single processor which takes 1 nanosecond per multiplication on average, roughly what is the total time taken per timestep (for the full 1 000 000 particles).
3. (a) Prove from the definition of a characteristic (see Section 8.3.2) that the equation of the characteristics of the collisionless Boltzmann equation is

$$\frac{d}{dt}\begin{pmatrix}\boldsymbol{x}\\ \boldsymbol{v}\end{pmatrix} = \begin{pmatrix}\boldsymbol{v}\\ \boldsymbol{a}\end{pmatrix}. \tag{10.11}$$

   Show also that a particle (or pseudo-particle) trajectory in a fixed electric potential depends only on initial velocity and the ratio of its charge to mass q/m, and therefore that the equation of motion of a pseudo-particle must use the same q/m as the actual particles, if a PIC simulation is to model correctly the force law at given speed and potential.

   (b) A pseudo-particle of charge q follows a characteristic, but it is supposed to be representative of many nearby particles (characteristics). If the mean density of pseudo-particles in a PIC simulation is a factor $1/g$ (where $g \gg 1$) smaller than the actual density of the system being modeled, how much charge must each pseudo-particle deposit on the potential grid to give the correct potential from Poisson's equation? One way to do PIC simulation is to represent all lengths, times, charges, and masses in physical units, but to use this charge deposition factor, and correspondingly lower particle density.

11
Monte Carlo techniques

So far we have been focussing on how particle codes work once the particles are launched. We've talked about how they are moved, and how self-consistent forces on them are calculated. What we have not addressed is how they are launched in an appropriate way in the first place, and how particles are reinjected into a simulation. We've also not explained how one decides statistically whether a collision has taken place to any particle and how one would then decide what scattering angle the collision corresponds to. All of this must be determined in computational physics and engineering by the use of random numbers and statistical distributions.[1] Techniques based on random numbers are called by the name of the famous casino at Monte Carlo.

11.1 Probability and statistics

11.1.1 Probability and probability distribution

Probability, in the mathematically precise sense, is an idealization of the repetition of a measurement, or a sample, or some other test. The result in each individual case is supposed to be unpredictable to some extent, but the repeated tests show some average trends that it is the job of probability to represent. So, for example, the single toss of a coin gives an unpredictable result: heads or tails; but the repeated toss of a (fair) coin gives on average equal numbers of heads and tails. Probability theory describes that regularity by saying the probability of heads and tails is equal. Generally, the probability of a particular class of outcomes (e.g. heads) is defined as the *fraction of the outcomes*, in a very

[1] S. Brandt (2014), *Data Analysis Statistical and Computational Methods for Scientists and Engineers*, fourth edition, Springer, New York, gives a much more expansive introduction to statistics and Monte Carlo techniques.

large number of tests, that are in the particular class. For a fair coin toss, the probability of heads is the fraction of outcomes of a large number of tosses that is heads, 0.5. For a six-sided die, the probability of getting any particular value, say 1, is the fraction of rolls that come up 1, in a very large number of tests. That will be one-sixth for a fair die. In all cases, because probability is defined as a *fraction*, the sum of probabilities of all possible outcomes must be unity.

More usually, in describing physical systems we deal with a *continuous* real-valued outcome, such as the speed of a randomly chosen particle. In that case the probability is described by a "probability distribution" $p(v)$, which is a function of the random variable (in this case velocity v). The probability of finding that the velocity lies in the range $v \rightarrow v + dv$ for small dv is then equal to $p(v)dv$. In order for the sum of all possible probabilities to be unity, we require

$$\int p(v)dv = 1. \tag{11.1}$$

Each individual sample[2] might give rise to more than one value. For example the velocity of a sampled particle might be a three-dimensional vector $\boldsymbol{v} = (v_x, v_y, v_z)$. In that case, the probability distribution is a function in a multidimensional parameter-space, and the probability of obtaining a sample that happens to be in a multidimensional element d^3v at $\boldsymbol{v}$ is $p(\boldsymbol{v})d^3v$. The corresponding normalization is

$$\int p(\boldsymbol{v})d^3v = 1. \tag{11.2}$$

Obviously, what this shows is that if our sample consists of randomly selecting particles from a velocity distribution function $f(\boldsymbol{v})$, then the corresponding probability function is simply

$$p(\boldsymbol{v}) = f(\boldsymbol{v}) \Big/ \int f(\boldsymbol{v})d^3v = f(\boldsymbol{v})/n, \tag{11.3}$$

where n is the particle density. So the normalized distribution function is the velocity probability distribution.

The *cumulative* probability function can be considered to represent the probability that a sample value is less than a particular value. So for a single-parameter distribution $p(v)$, the cumulative probability is

$$P(v) = \int_{-\infty}^{v} p(v')dv'. \tag{11.4}$$

[2] Statisticians use the generic word "sample" to refer to the particular result of a single test or measurement.

In multiple dimensions, the cumulative probability is a multidimensional function that is the integral in all the dimensions of the probability distribution:

$$P(\boldsymbol{v}) = P(v_x, v_y, v_z) = \int_{-\infty}^{v_x} \int_{-\infty}^{v_y} \int_{-\infty}^{v_z} p(\boldsymbol{v}')d^3v'. \tag{11.5}$$

Correspondingly, the probability distribution is the derivative of the cumulative probability: $p(v) = dP/dv$, or $p(\boldsymbol{v}) = \partial^3 P/\partial v_x \partial v_y \partial v_z$.

11.1.2 Mean, variance, standard deviation, and standard error

If we make a large number N of individual measurements of a random value from a probability distribution $p(v)$, each of which gives a value v_i, $i = 1, 2, \ldots, N$, then the *sample mean* value of the combined sample N is defined as

$$\mu_N = \frac{1}{N} \sum_{i=1}^{N} v_i. \tag{11.6}$$

The *sample variance* is defined[3] as

$$S_N^2 = \frac{1}{N-1} \sum_{i=1}^{N} (v_i - \mu_N)^2. \tag{11.7}$$

The *sample standard deviation* is S_N, the square root of the variance, and the *sample standard error* is $S_N/\sqrt{N}$. The mean is obviously a measure of the average value, and the variance or standard deviation is a measure of how spread out the random values are. They are the simplest unbiassed estimates of the moments of the distribution. These moments are properties of the *probability distribution* not of the particular sample. The *distribution mean*[4] is defined as

$$\mu = \int v p(v) dv \tag{11.8}$$

and the *distribution variance* is

$$S^2 = \int (v - \mu)^2 p(v) dv. \tag{11.9}$$

[3] Division by the factor $N - 1$ rather than N makes this formula an unbiassed estimate of the distribution variance. One way to understand this is to recognize that the number of degrees of freedom of $\sum_1^N (v_i - \mu_N)^2$ is $N - 1$, not N. Using $N - 1$ is sometimes called "Bessel's correction".

[4] Often called the "expectation" of v.

Obviously for large N we expect the sample mean to be approximately equal to the distribution mean and the sample variance equal to the distribution variance.

A finite-size sample will not have a mean exactly equal to the distribution mean because of statistical fluctuations. If we regard the sample mean μ_N as itself being a random variable, which changes from one total sample of N tests to the next total sample of N tests, then it can be shown[5] that the probability distribution of μ_N is approximately a Gaussian with standard deviation equal to the standard error $S_N/\sqrt{N}$. That is one reason why the Gaussian distribution is sometimes called the "normal" distribution. The Gaussian probability distribution in one dimension has only two[6] independent parameters μ and S.

11.2 Computational random selection

Computers can generate *pseudo*-random numbers, usually by doing complicated non-linear arithmetic starting from a particular "seed," number (or strictly a seed "state," which might be multiple numbers). Each successive number produced is actually completely determined by the algorithm, but the sequence has the appearance of randomness, in that the values v jump around in the range $0 \leq v \leq 1$, with no apparent systematic trend to them. If the random-number generator is a good generator, then successive values will not have statistically detectable dependence on the prior values, and the

[5] **Enrichment: Central Limit Theorem.** It is not straightforward to prove that the distribution becomes Gaussian. But it is fairly easy to show that the variance of the sample mean is the variance of the distribution divided by N. From the definition of μ_N one can immediately deduce that

$$(\mu_N - \mu)^2 = \left(\frac{1}{N}\sum_1^N (v_i - \mu)\right)^2 = \frac{1}{N^2}\sum_{i,j}^N (v_i - \mu)(v_j - \mu).$$

Take the expectation $\langle \dots \rangle$ of this quantity to obtain the variance of the distribution of sample means:

$$\langle(\mu_N - \mu)^2\rangle = \frac{1}{N^2}\sum_{i,j}^N \langle(v_i - \mu)(v_j - \mu)\rangle = \frac{1}{N^2}\sum_i^N \langle(v_i - \mu)^2\rangle = \frac{S^2}{N}.$$

The first equality, taking the expectation inside the sum, is a simple property of taking the expectation: the expectation of a sum is the sum of the expectations. The second equality uses the fact that $\langle(v_i - \mu)(v_j - \mu)\rangle = 0$ for $i \neq j$ because the quantities $(v_i - \mu)$ are statistically independent and have zero mean. That is sufficient to yield the required result. Our estimate for the distribution variance is $S^2 = S_N^2$. So the unbiassed estimate for the variance of μ_N is $\langle(\mu_N - \mu)^2\rangle = S_N^2/N$. The standard error is the square root of this quantity.

[6] An un-normalized Gaussian distribution has three, including the height.

distribution of values in the range will be uniform, representing a probability distribution $p(v) = 1$. Many languages and mathematical systems have library functions that return a random-number. Not all such functions are "good" random number generators. (The built-in C functions are notoriously not good.) One should be wary for production work. It is also extremely useful, for example for program debugging, to be able to repeat a pseudo-random calculation, knowing the sequence of "random" numbers you get each time will be the same. What you must be careful about, though, is that if you want to improve the accuracy of a calculation by increasing the number of samples, it is essential that the samples be independent. Obviously, that means the random numbers you use must *not* be the same ones you already used. In other words, the seed must be different. This goes also for parallel computations. Different parallel processors should normally use different seeds.

Now obviously if our computational task calls for a random number from a uniform distribution between 0 and 1, $p(v) = 1$, then using one of the internal or external library functions is the way to go. However, usually we will be in a situation where the probability distribution we want to draw from is *non-uniform*, for example a Gaussian distribution, an exponential distribution, or some other function of value. How do we do that?

We use two related random variables; call them u and v. Variable u is going to be uniformly distributed between 0 and 1. (It is called a "uniform deviate".) Variable v is going to be related to u through some one-to-one functional relation. Now if we take a particular sample value drawn from the uniform deviate, u, there is a corresponding value v. What's more, we know that the fraction of drawn values that are in a particular u-element du is equal to the fraction of values that are in the corresponding v-element dv. Consequently, recognizing that those fractions are $p_u(u)du$ and $p_v(v)dv$, respectively, where p_u and p_v are the respective probability distributions of u and v, we have

$$p_u(u)du = p_v(v)dv \qquad \Rightarrow \qquad p_v(v) = p_u(u)\left|\frac{du}{dv}\right| = \left|\frac{du}{dv}\right|. \tag{11.10}$$

The final equality uses the fact that $p_u = 1$ for a uniform deviate.

Therefore, if we are required to find random values v from a probability distribution p_v, we simply have to find a functional relationship between v and u that satisfies $p_v(v) = |du/dv|$. But we know of a function already that provides this property. Consider the cumulative probability $P_v(v) = \int^v p_v(v')dv'$. It is monotonic, and ranges between 0 and 1. Therefore we may choose to write

$$u = P_v(v) \qquad \text{for which} \qquad \frac{du}{dv} = p_v(v). \tag{11.11}$$

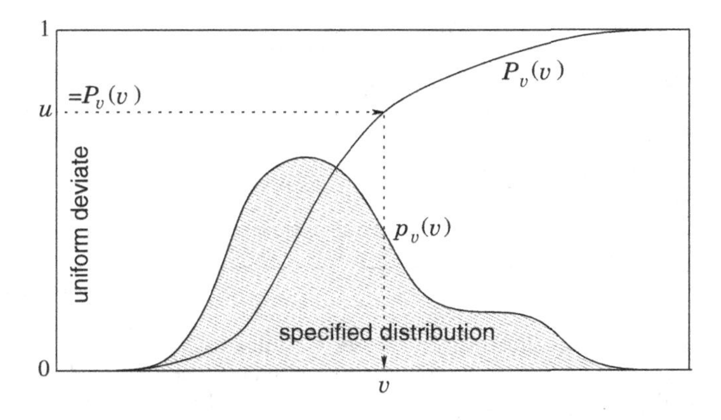

Figure 11.1 To obtain numerically a random variable v with specified probability distribution p_v (not to scale), calculate a table of the function $P_v(v)$ by integration. Draw a random number from uniform deviate u. Find the v for which $P_v(v) = u$ by interpolation. That's the random v.

So if $u = P_v(v)$, the v variable will be randomly distributed with probability distribution $p_v(v)$. We are done. Actually not quite done, because the process of choosing u and then finding the value of v which corresponds to it requires us to invert the function $P_v(v)$. That is

$$v = P_v^{-1}(u). \tag{11.12}$$

Figure 11.1 illustrates this process. It is not always possible to invert the function analytically, but it is always possible to do it numerically. One way is by root finding, e.g. bisection. Since P_v is monotonic, for any u between 0 and 1, there is a single root v of the equation $P_v(v) - u = 0$. Provided that we can find that root quickly, then given u we can find v. One way to make the root-finding quick is to generate a table of values of v and $u = P_v(v)$, of length N_t, equally spaced *in* u (not in v). Then, given any u, the index of the point just below u is the integer value $i = u{*}N_t$, and we can interpolate between it and the next point using the fractional value of $u{*}N_t$.[7]

Rejection method Another way of obtaining random values from some specified probability distribution is by the "rejection method", illustrated in Fig. 11.2. This involves using a second random number to decide whether or not to retain the first one chosen. The second random number is used to weight

[7] Linear interpolation is then equivalent to representing p_v as a histogram. So adequate resolution may require a fairly large number N_t.

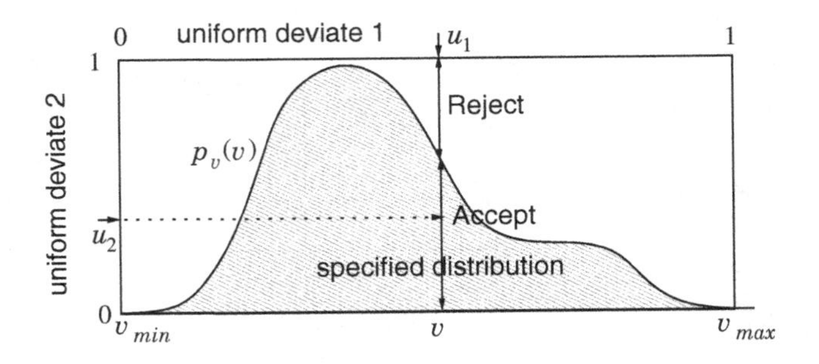

Figure 11.2 The rejection method chooses a v value randomly from a simple distribution (e.g. a constant) whose integral is invertible. Then a second random number decides whether it will be rejected or accepted. The fraction accepted at v is equal to the ratio of $p_v(v)$ to the simple invertible distribution. $p_v(v)$ must be scaled by a constant factor to be everywhere less than the simple distribution (1 here).

the probability of the first one. In effect this means picking points below the first scaled distribution, in the illustrated case of a rectangular distribution, uniformly distributed within the rectangle, and accepting only those that are below $p_v(v)$ (suitably scaled to be everywhere less than 1). Therefore some inefficiency is inevitable. If the area under $p_v(v)$ is, say, half the total, then twice as many total choices are needed, and each requires two random numbers, giving four times as many random numbers per accepted point. Improvement on the second inefficiency can be obtained by using a simply invertible function that fits $p_v(v)$ more closely. Even so, this will be slower than the tabulated function method, unless the random-number generator has very small cost.

Monte Carlo integration Notice, by the way, that this second technique shows exactly how "Monte Carlo integration" can be done. Select points at random over a line, or a rectangular area in two dimensions, or cuboid volume in three dimensions. Decide whether each point is within the area/volume of interest. If so, add the value of the function to be integrated to the running total, if not, not. Repeat. At the end multiply the total by the area/volume of the rectangle/cuboid divided by the number of random points examined (total, not just those that are within the area/volume). That's the integral. Such a technique can be quite an efficient way, and certainly an easy-to-program way, to integrate over a volume for which it is simple to decide whether you are inside it but hard to define systematically where the boundaries are. An example might be the volume inside a cube but outside a sphere placed off-center inside the cube. The method's drawback is that its accuracy increases only like the inverse *square root* of the number of points sampled. So, if high accuracy is required, other methods may be much more efficient.[8]

11.3 Flux integration and injection choice

Suppose we are simulating a subvolume that is embedded in a larger region. Particles move in and out of the subvolume. Something interesting is being modeled within the subvolume, for example the interaction of some object with the particles. If the volume is big enough, the region outside the subvolume is relatively unaffected by the physics in the subvolume, then we know or can specify what the distribution function of particles is in the outer region, at the volume's boundaries. Assume that periodic boundary conditions are not appropriate, because, for example, they don't well represent an isolated interaction. How do we determine statistically what particles to inject into the subvolume across its boundary?

Suppose the volume is a cuboid shown in Fig. 11.3. It has six faces, each of which is normal to one of the coordinate axes, and located at $\pm L_x$, $\pm L_y$ or $\pm L_z$. We'll consider the face perpendicular to x, which is at $-L_x$, so that positive velocity v_x corresponds to moving *into* the simulation subvolume. We calculate the rate at which particles are crossing the face into the subvolume. If the distribution function is $f(\boldsymbol{v}, \boldsymbol{x})$, then the flux density in the $+v_x$ direction is

$$\Gamma_x(\boldsymbol{x}) = \int\int\int_{v_x=0}^{\infty} v_x f(\boldsymbol{v}, \boldsymbol{x}) dv_x dv_y dv_z \tag{11.13}$$

and the number entering across the face per unit time (the flux) is

$$F_{-L_x} = \int_{-L_y}^{L_y}\int_{-L_y}^{L_y} \Gamma_x(-L_x, y, z)\, dydz. \tag{11.14}$$

[8] **Enrichment: Quiet start and quasi-random selection.** When starting a particle-in-cell (PIC) simulation, the initial positions of the particles might be chosen using random numbers to decide their location. However, they then will have density fluctuations of various wavelengths that in plasmas may be *bigger* than are present after running the simulation for many steps. The reason for this discrepancy is that the feedback effect of the self-consistent electric potential tends to smooth out density fluctuations, so that in the fully developed simulation the noise level is lower than purely random. A simple way of saying the same thing is that individual particles *repel* others of the same type, preventing clumping of the particles. It is therefore often physically reasonable to start a PIC simulation with positions that are chosen to be more evenly spaced than purely random. Indeed, for some calculations it is advantageous (but non-physical) to start with density fluctuations even *lower* than the final level that would be present for a steady plasma. In either case, what is called a "quiet start" can be obtained by using what are called "quasi-random" numbers (see, e.g., *Numerical Recipes*) instead of (pseudo) random numbers. Quasi-random numbers are somewhat random, but much smoother in their distribution because each new number takes account of the already used numbers and tries to avoid being close to them. Successive numbers are thus correlated rather than uncorrelated. For Monte Carlo integration, such smoother distributions in space are also often highly appropriate and can give lower-noise results for the same number of samples, beating the weak, $1/\sqrt{N}$, decrease of fractional error.

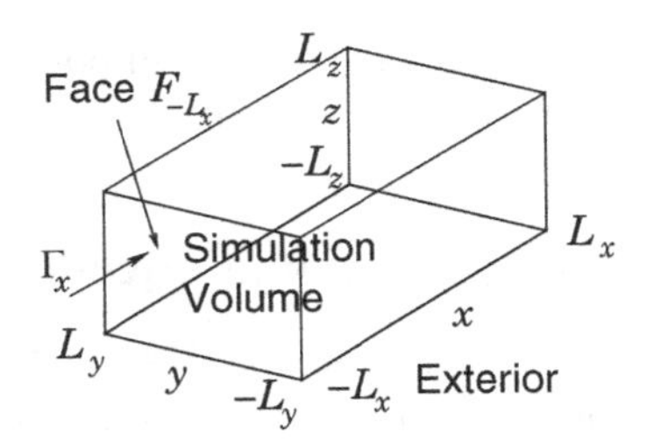

Figure 11.3 Simulating over a volume that is embedded in a wider external region, we need to be able to decide how to inject particles from the exterior into the simulation volume so as to represent statistically the exterior distribution.

Assume that the outer region is steady, independent of time. We proceed by evaluating the six fluxes F_j, using the above expressions and their equivalents for the six different faces. At each timestep of the simulation, we decide how many particles to inject into the subvolume from each face. The average number of particles to inject in a timestep Δt at face j is equal to $F_j \Delta t$. If this number is large,[9] then it may be appropriate to inject just that number (although dealing with the non-integer fractional part of the number needs consideration). But if it is of order unity or even smaller, then that does not correctly represent the statistics. In that case we need to decide statistically at each step how many particles are injected: 0, 1, 2 . . . .

It is a standard result of probability theory[10] that if events (in this case injections) take place randomly and uncorrelated with each other at a fixed

[9] Or if we wish to do "quiet injection" that is smoother than purely random.

[10] **Enrichment: The discrete Poisson distribution.** Suppose there is a very large number N of similar, uncorrelated, events (for example radioactive decays of atoms) waiting to occur. In a time duration far shorter than the waiting time (e.g. the half-life) of an individual event the probability that any one of them occurs is small, say $p = r/N$. Then r is the average total number of events occuring in this time duration. The total number of events actually occuring in a particular time sample is then an integer, and we want to find the probability of each possible integer. Any sample consists of N choices about the individual events: yes or no. The number of yes events is distributed as a binomial distribution, in which the probability of n yes events is given by the number of different ways to choose n out of N, which is $\frac{N!}{n!(N-n)!}$, times the probability of each specific arrangement of n yes events and $N-n$ no events, $p^n(1-p)^{N-n}$. The total is

$$p_n = \frac{N!}{n!(N-n)!} p^n (1-p)^{N-n} = \frac{r^n}{n!}\left(1 - \frac{r}{N}\right)^N \left[\frac{1}{N^n(1-r/N)^n} \frac{N!}{(N-n)!}\right].$$

Now recognize that the limit for large N (but constant n) of the square bracket term is 1; while the limit of the term $\left(1 - \frac{r}{N}\right)^N$ is $\exp(-r)$. Therefore the probability of obtaining n total events, of a type that are completely uncorrelated ($N \to \infty$), when their average rate of occurrence is r, is

$$p_n = \frac{r^n}{n!} \exp(-r).$$

This is the discrete Poisson distribution.

average rate r (per sample, in this case per timestep) then the number n that happens in any particular sample is an integer random variable with "Poisson distribution": a discrete probability distribution

$$p_n = \exp(-r)r^n/n!\,. \tag{11.15}$$

The parameter giving the rate, r, is a real number, but the number for each sample, n, is an integer. One can rapidly verify that, since $\sum_{n=0}^{\infty} r^n/n! = \exp(r)$, the probabilities are properly normalized: $\sum_n p_n = 1$. The mean rate is $\sum_n np_n = r$ (as advertized). The variance, it turns out, is also r. So the standard deviation is $\sqrt{r}$. The value p_n gives us precisely the probability that n particles will need to be injected when the flux is $r = F_j$. So the first step in deciding injections is to select randomly a number to be injected, from the Poisson distribution eq. (11.15). There are various ways to do this (including library routines). The root-finding approach is easily applied, because the cumulative probability function $P_u(u)$ can be considered to consist of steps of height p_n at the integers n (and constant in between).

Next, we need to decide where on the surface each injection is going to take place. If the flux density is uniform, then we just pick randomly a position corresponding to $-L_y \le y \le L_y$ and $-L_z \le z \le L_z$. Non-uniform flux density, however, introduces another distribution function inversion headache. It's more work, but straightforward.

Finally, we need to select the actual velocity of the particle. Very importantly, the probability distribution of this selection is *not* just the velocity distribution function, or the velocity distribution function restricted to positive v_x. No, it is the *flux* distribution $v_x f(\boldsymbol{v},\boldsymbol{x})$ weighted by the normal velocity v_x (for an x-surface) if positive (otherwise zero). If the distribution is separable, $f(\boldsymbol{v}) = f_x(v_x)f_y(v_y)f_z(v_z)$, as it is if it is Maxwellian, then the tangential velocities v_y, v_z, can be treated separately: select two different velocities from the respective distributions. And select v_x (positive) from a probability distribution proportional to $v_x f_x(v_x)$.

If f is not separable, then a more elaborate random selection is required. Suppose we have the cumulative probability distribution of velocities $P(v_x, v_y, v_z)$ for all interesting values of velocity. Notice that if v_{xmax}, v_{ymax}, v_{zmax} denote the largest relevant values of v_x, v_y and v_z, beyond which $f = 0$, then $P(v_x, v_{ymax}, v_{zmax})$ is a function of just one variable v_x, and is the cumulative probability distribution integrated over the entire relevant range of the other velocity components. That is, it is the one-dimensional cumulative probability distribution of v_x. We can convert it into a one-dimensional flux cumulative probability by performing the following integral (numerically, discretely):

$$F(v_x) = \int_0^{v_x} v'_x \frac{\partial}{\partial v'_x} P(v'_x, v_{ymax}, v_{zmax})\, dv'_x \tag{11.16}$$
$$= v_x P(v_x, v_{ymax}, v_{zmax}) - \int_0^{v_x} P(v'_x, v_{ymax}, v_{zmax})\, dv'_x.$$

Afterwards, we can normalize $F(v_x)$ by dividing by $F(v_{xmax})$, ariving at the cumulative flux-weighted probability for v_x.

We then proceed as follows.

1. Choose a random v_x from its cumulative flux-weighted probability $F(v_x)$.
2. Choose a random v_y from its cumulative probability for the already chosen v_x, namely $P(v_x, v_y, v_{zmax})$ regarded as a function only of v_y.
3. Choose a random v_z from its cumulative probability for the already chosen v_x and v_y, namely $P(v_x, v_y, v_z)$ regarded as a function only of v_z.

Naturally for other faces y and z one has to start with the corresponding velocity component and cycle round the indices. For steady external conditions all the cumulative velocity probabilities need to be calculated only once, and then stored for subsequent timesteps.

Discrete-particle representation An alternative to this continuum approach is to suppose that the external distribution function is represented by a perhaps large number, N, (maybe millions) of representative "particles" distributed in accordance with the external velocity distribution function. Particle k has velocity $\boldsymbol{v}_k$ and the density of particles in phase-space is proportional to the distribution function, that is to the probability distribution. Then if we wished to select randomly a velocity from the particle distribution we simply arrange the particles in order and pick one of them at random. However, when we want the particles to be flux-weighted, in normal direction $\hat{\boldsymbol{n}}$ say, we must select them with probability proportional to $v_n = \hat{\boldsymbol{n}}.\boldsymbol{v}$ (when positive, and zero when negative). Therefore, for this normal direction we must consider each particle to have appropriate weight. We associate each particle with a real[11] index r so that when $r_k \leq r < r_{k+1}$ the particle k is indicated. The interval length allocated to particle k is chosen proportional to its weight, so that $r_{k+1} - r_k \propto \hat{\boldsymbol{n}}.\boldsymbol{v}_k$. Then the selection consists of a random-number draw x, multiplied by the total real index range and indexed to the particle and hence to its velocity: $x(r_{N+1} - r_1) + r_1 = r \rightarrow k \rightarrow \boldsymbol{v}_k$. The discreteness of the particle distribution will generally not be an important limitation for a process that already relies on random particle representation. The position of injection

[11] Or double-precision if N is very large.

will anyway be selected differently even if a representative particle velocity is selected more than once.

Worked example. High-dimensionality integration

Monte Carlo techniques are commonly used for high-dimensional problems; integration is perhaps the simplest example. The reasoning is approximately this. When there are d dimensions, the total number of points in a grid whose size is M in each coordinate direction is $N = M^d$. The fractional uncertainty in estimating a one-dimensional integral of a function with only isolated discontinuities, based upon evaluating it at M grid points, may be estimated as $\propto 1/M$. If this estimate still applies to multiple-dimension integration (and this is the dubious part), then the fractional uncertainty is $\propto 1/M = N^{-1/d}$. By comparison, the uncertainty in a Monte Carlo estimate of the integral based on N evaluations is $\propto N^{-1/2}$. When d is larger than 2, the Monte Carlo square-root convergence scaling is better than the grid estimate. And if d is very large, Monte Carlo is much better. Is this argument correct? Test it by obtaining the volume of a four-dimensional hypersphere by numerical integration, comparing a grid technique with Monte Carlo.

A four-dimensional sphere of radius 1 consists of all those points for which $r^2 = x_1^2 + x_2^2 + x_3^2 + x_4^2 \leq 1$. Its volume is known analytically; it is $\pi^2/2$. Let us evaluate the volume numerically by examing the unit hypercube $0 \leq x_i \leq 1$, $i = 1, \ldots, 4$. It is $1/2^4 = 1/16$th of the hypercube $-1 \leq x_i \leq 1$, inside of which the hypersphere fits; so the volume of the hypersphere that lies within the $0 \leq x_i \leq 1$ unit hypercube is $1/16$th of its total volume; it is $\pi^2/32$. We calculate this volume numerically by discrete integration as follows.

A deterministic (non-random) integration of the volume consists of constructing an equally spaced lattice of points at the center of cells that fill the unit cube. If there are M points per edge, then the lattice positions in the dimension $i\,(i = 1, \ldots, 4)$ of the cell-centers are $x_{i,k_i} = (k_i - 0.5)/M$, where $k_i = 1, \ldots, M$ is the (dimension-i) position index. We integrate the volume of the sphere by collecting values from every lattice point throughout the unit hypercube. A value is unity if the point lies within the hypersphere $r^2 \leq 1$; otherwise it is zero. We sum all values (0 or 1) from every lattice point and obtain an integer equal to the number of lattice points S inside the hypersphere. The total number of lattice points is equal to M^4. That sum corresponds to the total volume of the hypercube, which is 1. Therefore the discrete estimate of the volume of $1/16$th of the hypersphere is S/M^4. We can compare this numerical integration with the analytic value and express the fractional error as the numerical value divided by the analytic value, minus one:

$$\text{Fractional error} = \left| \frac{S/M^4}{\pi^2/32} - 1 \right| .$$

Monte Carlo integration works essentially exactly the same except that the points we choose are not a regular lattice, but rather they are random. Each one is found by taking four new uniform-variate values (between 0 and 1) for the four coordinate values x_i. The point contributes unity if it has $r^2 \leq 1$ and zero otherwise. We obtain a different count S_r. We'll choose to use a total number N of random point positions exactly equal to the number of lattice points $N = M^4$, although we could have made N any integer we liked. The Monte Carlo integration estimate of the volume is S_r/N.

I wrote a computer code to carry out these simple procedures, and compare the fractional errors for values of M ranging from 1 to 100. The results are shown in Fig. 11.4.

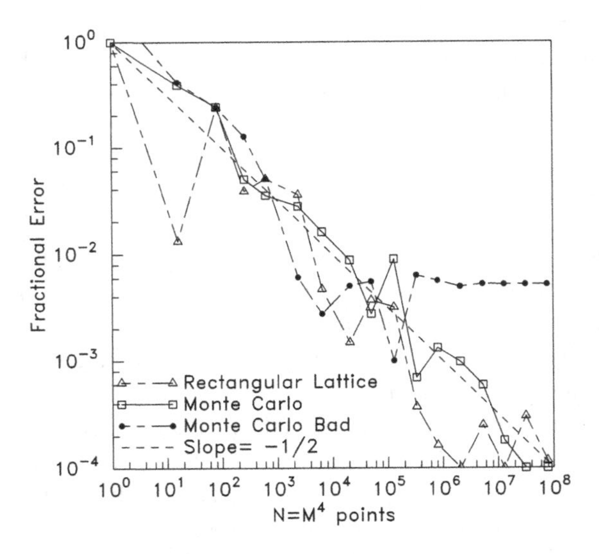

Figure 11.4 Comparing error in the volume of a hypersphere found numerically using lattice and Monte Carlo integration. It turns out that Monte Carlo integration actually does *not* converge significantly faster than lattice integration, contrary to common wisdom. They both converge approximately like $1/\sqrt{N}$ (logarithmic slope $= -1/2$). What's more, if one uses a "bad" random-number generator (the Monte Carlo Bad line) it is possible that the random integration will cease converging at some number, because it gives only a finite length of independent random numbers, which in this case is exhausted at roughly a million.

Four-dimensional lattice integration does as well as Monte Carlo for this sphere. Lattice integration is not as bad as the dubious assumption of fractional uncertainty $1/M = N^{-1/d}$ suggests; it is more like $N^{-2/d}$ for $d > 1$. Only at higher dimensionality than $d = 4$ do tests show the advantages of Monte Carlo integration beginning to be significant. As a bonus, this integration experiment detects poor random-number generators.[12]

Exercise 11. Monte Carlo techniques

1. A random variable is required, distributed on the interval $0 \leq x \leq \infty$ with probability distribution $p(x) = k\exp(-kx)$, with k a constant. A library routine is available that returns a uniform random variate y (i.e. with uniform probability $0 \leq y \leq 1$). Give formulas and an algorithm to obtain the required randomly distributed x value from the returned y value.
2. Particles that have a Maxwellian distribution

$$f(\mathbf{v}) = n\left(\frac{m}{2\pi kT}\right)^{3/2} \exp\left(-\frac{mv^2}{2kT}\right) \tag{11.17}$$

   cross a boundary into a simulation region.

   (a) Find the cumulative probability distribution $P(v_n)$ that ought to be used to determine the velocity v_n normal to the boundary; of the injected particles.
   (b) What is the total rate of injection per unit area that should be used?
   (c) If the timestep duration is Δt, and the total rate of crossing a face is r such that $r\Delta t = 1$, what probability distribution should be used to determine statistically the *actual* integer number of particles injected at each step?
   (d) What should be the probability of 0, 1, or 2 injections?
3. Write a code to perform a Monto Carlo integration of the area under the curve $y = (1 - x^2)^{0.3}$ for the interval $-1 \leq x \leq 1$. Experiment with different total numbers of sample points, and determine the area accurate to 0.2%, and approximately how many sample points are needed for that accuracy.

[12] The random generators I used here are both portable. The good generator is the `RANLUX` routine described by M. Luscher, (1994), *Computer Physics Communications* **79** 100, and F. James, (1994), *Computer Physics Communications* **79** 111, used with the lowest luxury level. The bad generator is the `RAN1` routine from the first (FORTRAN) edition of *Numerical Recipes*. It was replaced by better routines in later editions.

12

Monte Carlo radiation transport

12.1 Transport and collisions

Consider the passage of uncharged particles through matter. The particles might be neutrons, or photons such as gamma rays. The matter might be solid, liquid, or gas, and contain multiple species with which the particles can interact in different ways. We might be interested in the penetration of the particles into the matter from the source, for example what is the particle flux at a certain place, and we might want to know the extent to which certain types of interaction with the matter have taken place, for example radiation damage or ionization. This sort of problem lends itself to modelling by Monte Carlo methods.

Since the particles are uncharged, they travel in straight lines at constant speed between collisions with the matter. Actually, the technique can be generalized to treat particles that experience significant forces so that their tracks are curved. However, charged particles generally experience many collisions that perturb their velocity only very slightly. Those small-angle collisions are not so easily or efficiently treated by Monte Carlo techniques, so we simplify the treatment by ignoring particle acceleration between collisions.

A particle executes a random walk through the matter, illustrated in Fig. 12.1. It travels a certain distance in a straight line, then collides. After the collision it has a different direction and speed. It takes another step in the new direction, generally with a different distance, to the next collision. Eventually, the particle has an absorbing collision, or leaves the system, or becomes so degraded (for example in energy) that it need no longer be tracked. The walk ends.

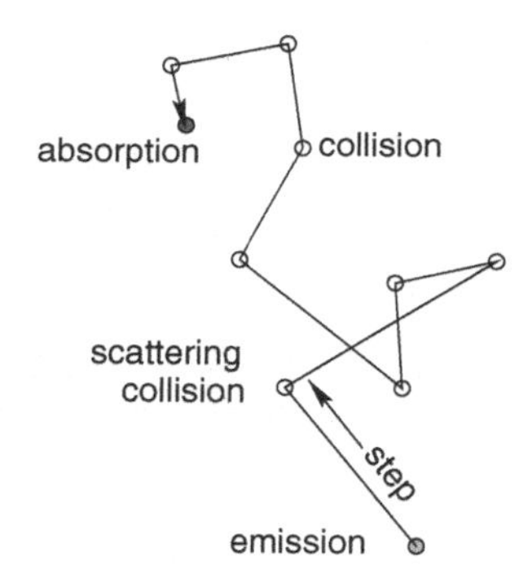

Figure 12.1 A random walk is executed by a particle with variable-length steps between individual scatterings or collisions. Eventually, it is permanently absorbed. The statistical distribution of the angle of scattering is determined by the details of the collision process.

12.1.1 Random-walk step length

The length of any of the steps between collisions is random. For any collision process, the average number of collisions a particle has per unit length corresponding to that process, which we'll label j, is $n_j\sigma_j$, where σ_j is the cross-section, and n_j is the density in the matter of the target species that has collisions of type j. For example, if the particles are neutrons and the process is uranium nuclear fission, n_j would be the density of uranium nuclei. The average total number of collisions of all types per unit path length is therefore

$$\Sigma_t = \sum_j n_j\sigma_j, \tag{12.1}$$

where the sum is over all possible collision processes. Once again Σ_t is an inverse attenuation-length.

How far does a particle go before having a collision? Well, even for particles with identical position and velocity, it varies statistically in an unpredictable way. However, the probability of having a collision in the small distance $l \to l + dl$, given that the particle started at l (in other words, earlier collisions are excluded), is $\Sigma_t dl$. This incremental collision probability $\Sigma_t dl$ is independent of prior events. Therefore (refer to Fig. 12.2), if the probability that the particle survives at least as far as l *without* a collision is $\bar{P}(l)$, then the probability that it survives at least as far as $l + dl$ is the product of $\bar{P}(l)$ times the probability $1 - \Sigma_t dl$ that it *does not* have a collision in dl:

$$\bar{P}(l + dl) = \bar{P}(l)(1 - \Sigma_t dl). \tag{12.2}$$

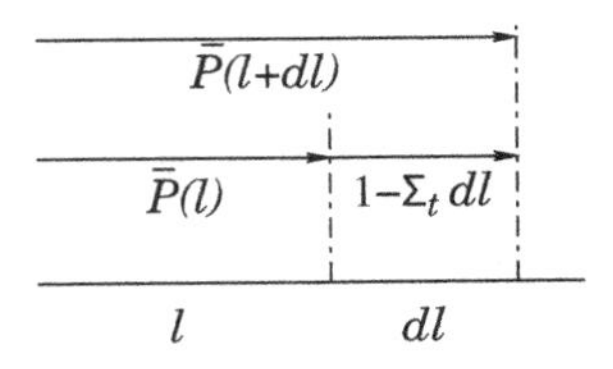

Figure 12.2 The probability, $\bar{P}(l)$, of survival without a collision to position l decreases by a multiplicative factor $1 - \Sigma_t dl$ in an increment dl.

Hence,

$$\frac{\bar{P}(l+dl) - \bar{P}(l)}{dl} \underset{dl \to 0}{\longrightarrow} \frac{d\bar{P}}{dl} = -\Sigma_t \bar{P}. \tag{12.3}$$

The solution of this differential equation is

$$\bar{P}(l) = \exp(-\int_0^l \Sigma_t dl), \tag{12.4}$$

where l is measured from the starting position, so $\bar{P}(0) = 1$. Another equivalent way to define $\bar{P}(l)$ is: the probability that the first collision is *not* in the interval $0 \to l$. The complement $P(l) = 1 - \bar{P}(l)$ is therefore the probability that the first collision *is* in $0 \to l$. In other words, $P(l) = 1 - \bar{P}(l)$ is a cumulative probability. It is the integral of the probability distribution $p(l)dl$ that the (first) collision takes place in dl at l; so:

$$p(l) = \frac{dP}{dl} = -\frac{d\bar{P}}{dl} = \Sigma_t \bar{P} = \Sigma_t \exp(-\int_0^l \Sigma_t dl). \tag{12.5}$$

To decide statistically where the first collision takes place for any specific particle, therefore, we simply draw a random number (uniform variate) and select l from the cumulative distribution $P(l)$ as explained in Section 11.2 and illustrated in Fig. 11.1. If we are dealing with a step in which Σ_t can be taken as uniform, then $P(l) = 1 - \exp(-\Sigma_t l)$, and the cumulative function can be inverted analytically as $l(P) = -\ln(1-P)/\Sigma_t$.

12.1.2 Collision type and parameters

Having decided where the next collision happens, we need to decide what sort of collision it is. The local rate at which each type j of collision is occurring is $n_j\sigma_j$ and the sum of all $n_j\sigma_j$ is Σ_t. Therefore, the fraction of collisions that is of type j is $n_j\sigma_j/\Sigma_t$. If we regard all the possible types of collisions as arrayed in a long list, illustrated in Fig. 12.3, and to each type is assigned a real number

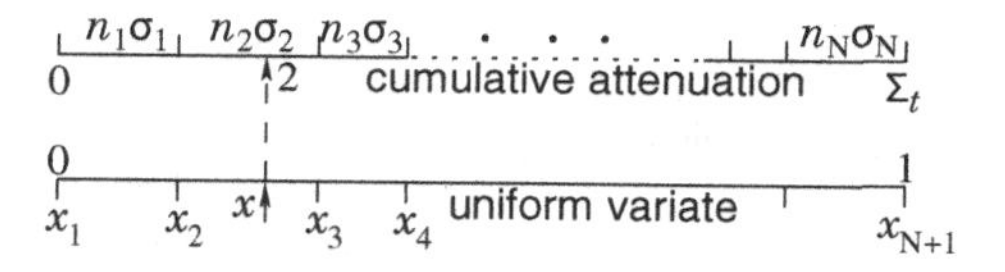

Figure 12.3 Deciding the collision type based upon a random number x.

x_j such that $x_{j+1} = x_j + n_j\sigma_j/\Sigma_t$, i.e.

$$x_j = \sum_{i=1}^{j-1} n_i\sigma_i/\Sigma_t, \tag{12.6}$$

then a single random draw x determines j by the condition $x_j \leq x < x_{j+1}$. This is just the process of drawing from a discrete probability distribution. Our previous examples were the Poisson distribution, and the weighting of discrete particles for flux injection.

Once having decided upon the collision process j, there are (usually) other parameters of the collision to be decided. For example, if the collision is a scattering event, what is the scattering angle and the corresponding new scattered velocity $\boldsymbol{v}$ which serves[1] to determine the starting condition of the next step? These random parameters are also determined by statistical choice from the appropriate probability distributions, for example the normalized differential cross-section per unit scattering angle.

12.1.3 Iteration and new particles

Once the collision parameters have been decided, unless the event was an absorption, the particle is restarted from the position of collision with the new velocity and the next step is calculated. If the particle is absorbed, or escapes from the modeled region, then instead we start a completely new particle. Where and how it is started depends upon the sort of radiation transport problem we are solving. If it is transport from a localized known source, then that determines the initialization position, and its direction is appropriately chosen. If the matter is emitting, new particles can be generated, spread throughout the volume. Spontaneous emission, arising independently of the level of radiation in the medium, is simply a distributed source. It serves to determine the distribution of brand new particles, which are launched once all active particles have been followed till absorbed or escaped. Examples

[1] For photons, the "velocity" should be interpreted as a combination of energy (frequency or wavelength) and propagation direction, since the particle's speed is always the speed of light.

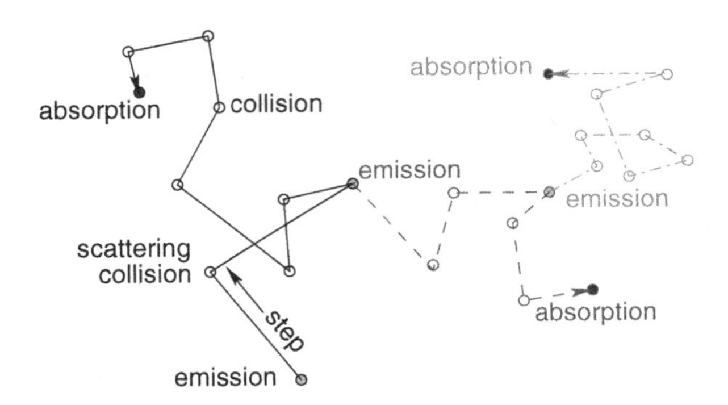

Figure 12.4 "Stimulated" emission is a source of new particles caused by the collisions of the old ones. It can multiply the particles through succeeding generations, as the new particles (different line styles) give rise themselves to additional emissions.

of spontaneous emission include radiation from excited atoms, spontaneous radioactive decays giving gammas, or spontaneous fissions giving neutrons. However, often the new particles are "stimulated" by collisions from the transporting particles. Fig. 12.4 illustrates this process. The classic nuclear example is induced fission producing additional neutrons. Stimulated emission arises as a product of a certain type of collision.

Any additional particles produced by collisions must be tracked by the same process as used for the original particle. They become new tracks. In a serial code, the secondary particles must be initialized and then set aside during the collision step. Then, when the original track ends by absorption or escape, the code must take up the job of tracking the secondary particles. The secondary particles may generate tertiary particles, which will also eventually be tracked. And so on. Parallel code might offload the secondary and tertiary tracks to other processors (or computational threads).

One might be concerned that we'll never get to the end if particles are generated faster than they are lost. That's true; we won't. But particle generation faster than loss corresponds to a runaway physical situation, for example a prompt supercritical reactor. So the real world will have problems much more troublesome than our computational ones!

12.2 Tracking, tallying, and statistical uncertainty

The purpose of a Monte Carlo simulation is generally to determine some averaged bulk parameters of the radiation particles or the matter itself.

The sorts of parameters include: the radiation flux as a function of position, the spectrum of particle energies, the rate of occurrence of certain types of collision, or the resulting state of the matter through which the radiation passes. To determine these sorts of quantities requires keeping track of the passage of particles through the relevant regions of space, and of the collisions that occur there. The computational task of keeping track of events and contributions to statistical quantities is often called "tallying."

What is generally done is to divide the region of interest into a managable number of discrete volumes, in which the tallies of interest are going to be accumulated. Each time an event occurs in one of these volumes, it is added to the tally. Then, provided a reasonably large number of such events has occurred, the average rate of the occurrence of events of interest can be obtained. For example, if we wish to know the fission power distribution in a nuclear reactor, we would tally the number of fission collisions occurring in each volume. Fission reactions release on average a known amount of total energy $\mathcal{E}$. So if the number occuring in a particular volume V during a time T is N, the fission power density is $N\mathcal{E}/VT$.

The number of computational random walks that we can afford to track is usually far less than the number of events that would actually happen in the physical system being modelled. Each computational particle can be considered to represent a large number of actual particles all doing the same thing. The much smaller computational number leads to important statistical fluctuations, uncertainty, or noise, that is a purely computational limitation.

The statistical uncertainty of a Monte Carlo calculation is generally determined by the observation that a sample of N random choices from a distribution (population) of standard deviation S has a sample mean μ_N whose standard deviation is equal to the standard error $S/\sqrt{N}$. Each tally event acts like single random choice. Therefore, for N tally events the uncertainty in the quantity to be determined is smaller than its intrinsic uncertainty or statistical range by a factor $1/\sqrt{N}$. Put differently, suppose the average rate of a certain type of statistically determined event is constant, giving on average N events in time T then the number n of events in any particular time interval of length T is governed by a Poisson probability, eq. (11.15) $p(n) = \exp(-N)N^n/n!$. The standard deviation of this Poisson distribution is $\sqrt{N}$. Therefore we obtain a precise estimate of the average rate of events, N, by using the number actually observed in a particular case, n, only if N (and therefore n) is a large number.

The first conclusion is that we must not choose our discrete tallying volumes too small. The smaller they are, the fewer events will occur in them, and, therefore, the less statistically accurate will be the estimate of the event rate.

When tallying collisions, the first thought one might have is simply to put a check mark against each type of collision for each volume element, whenever that exact type of event occurs in it. The difficulty with this approach is that it will give very poor statistics, because there are going to be too few check marks. To get, say, 1% statistical uncertainty, we would need 10^4 check marks for each discrete volume. If the events are spread over, say 100^3 volumes, we would need 10^{10} total collisions *of each relevant type*. For, say, 100 collision types we are already up to 10^{12} collisions. The cost is very great. But we can do better than this naïve check-box approach.

One thing we can do with negligible extra effort is to add to the tally for *every* type of collision, whenever *any* collision happens. To make this work, we must add a variable amount to the tally, not just 1 (not just a check mark). The amount we should add to each collision-type tally is just the fraction of all collisions that is of that type, namely $n_j\sigma_j/\Sigma_t$. Of course these fractional values should be evaluated with values of n_j corresponding to the local volume, and at the velocity of the particle being considered (which affects σ_j). This process will produce, on average, the same correct collision tally value, but will do so with a total number of contributions bigger by a factor equal to the number of collision types. That substantially improves the statistics.

A further statistical advantage is obtained by tallying more than just collisions. If we want particle flux properties, as well as the effects on the matter, we need to tally not just collisions, but also the average flux density in all the discrete volumes. That requires us to determine for every volume, every passage i of a particle through it, the fractional time Δt_i that it spends in the volume, and the speed v_i at which it passed through. See Fig. 12.5. When the simulation has been tracked for a sufficient number of particles, the density of particles in the volume is proportional to the sum $\sum_i \Delta t_i$ and the scalar

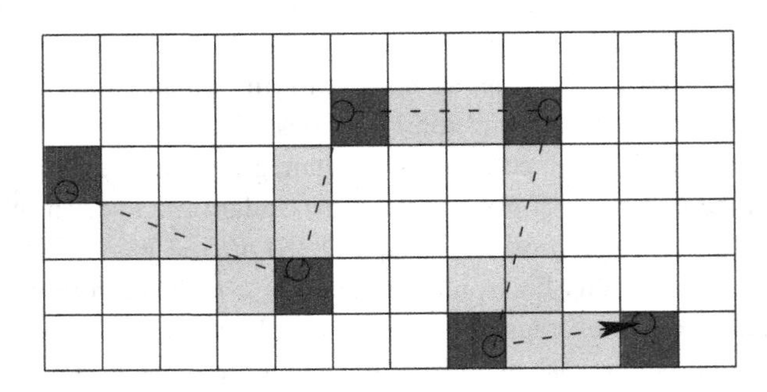

Figure 12.5 Tallying is done in discrete volumes. One can tally every collision but if the steps are larger than the volumes, it gives better statistics to tally every volume the particle passes through.

flux density[2] to $\sum_i v_i \Delta t_i$. If collisional lengths (the length of the random walk step) are substantially larger than the size of the discrete volumes, then there are more contributions to the flux property tallies than there are walk steps, i.e. modeled collisions. The statistical accuracy of the flux density measurement is then better than the collision tallies (even when all collision types are tallied for every collision) by a factor approximately equal to the ratio of collision step length to discrete volume side-length. Therefore, it may be worth the effort to keep a tally of the total contribution to collisions of type j that we care about from every track that passes through every volume. In other words, for each volume, to obtain the sum over all passages i: $\sum_i \Delta t_i v_i n_j \sigma_j$. The extra cost of this process is the geometric task of determining the length of a line during which it is inside a particular volume. But if that task is necessary anyway, because flux properties are desired, then it is no more work to tally the collisional probabilities in this way.

Importance weighting Another aspect of statistical accuracy relates to the choice of initial distribution of particles, especially in velocity-space. In some transport problems, it might be that the particles that penetrate a large distance away from their source are predominantly the high-energy particles, perhaps those quite far out on the tail of a birth velocity distribution that has the bulk of its particles at low energy. The most straightforward way to represent the transport problem is to launch particles in proportion to their physical birth distribution. Each computational particle then represents the same number of physical particles. But this choice might mean that there are very few of the high-energy particles that determine the flux far from the source. If so, then the statistical uncertainty of the far flux is going to be great. It is possible to compensate for this problem by *weighting* the particles, illustrated by Fig. 12.6. One can obtain better far-flux statistics by launching more of the high-energy particles than would be implied by their probability distribution, but compensating for the statistical excess by an importance weighting w that is inversely proportional to the factor by which the particle number has exceeded the probability distribution. For a weighted particle's subsequent random walk, and for any additional particles emitted in response to collisions of the particle, the contribution made to any flux or reaction rate is then multiplied by w. Consequently, the total flux calculated anywhere has the correct average contribution from all types of initial particles. But there are

[2] The velocity distribution f_k in discrete velocity bins $d^3 v_k$ may also be desired. It is given by summing only those passages that occur in the velocity bin: $f_k d^3 v_k = \sum_{v_i \in d^3 v_k} \Delta t_i$. It will have greater statistical uncertainty because of fewer samples per bin.

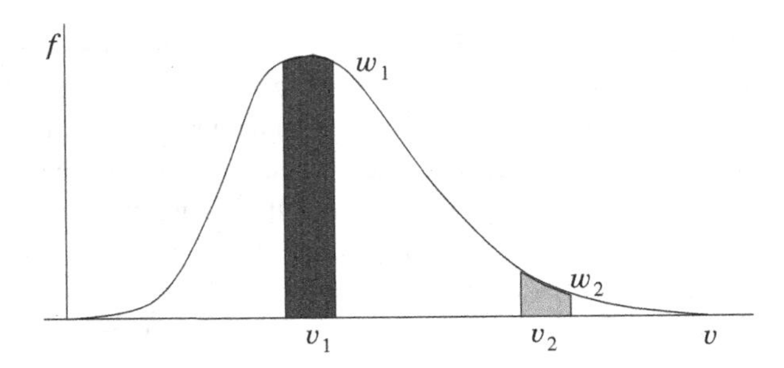

Figure 12.6 If all particles have the same weight, there are many more representing bulk velocities (v_1) than tail velocities (v_2). Improved statistical representation of the tail is obtained by giving the tail velocities lower weight w_2. If the weights are proportional to f then equal numbers of particles are assigned to equal velocity ranges.

more actual contributions (of lower weight) from high-initial-energy particles than there otherwise would be. This has the effect of reducing the statistical uncertainty of flux that happens to depend upon high-energy birth particles by transferring some of the computational effort to them from lower-energy birth particles.

Importance weighting is immediately accommodated during the tallying process by adding contributions to the various flux tallies that are the same as have been previously given – except that each contribution is multiplied by the weight w_i of that particular particle. Thus, density becomes $\propto \Sigma_i w_i \Delta t_i$ and scalar flux density $\propto \Sigma_i w_i v_i \Delta t_i$, and so on.

The weight of a particle that is launched as a result of stimulated emission is inherited from its parent. A newly launched particle of the nth generation inherently acquires the weight w_{n-1} of the particular previous-generation particle whose collision caused the launch. However, in the process of deciding the nth generation's initial velocity an extra launch weighting is accorded to it, corresponding to its launch parameters; for example, it might be weighted proportional to the launch-speed (or energy) probability distribution $p(v_n)$. If so, then the weight of the newly launched particle is

$$w_n = w_{n-1} p(v_n). \tag{12.7}$$

This process of weight multiplication between generations causes a *spreading of the weights*. The ratio of the maximum to minimum weight of each succeeding generation is multiplied by the birth weight ratio; so the weight range ratio grows geometrically. When there is a very large range between the minimum and the maximum weight, it becomes a waste of time to track particles whose weight is negligibly small. So after the simulation has run for some time duration, special procedures must be applied to limit the range of weights by removing particles of irrelevantly small weight.

Worked Example. Density and flux tallies

In a Monte Carlo calculation of radiation transport from a source emitting at a fixed rate R particles per second, tallies in the surrounding region are kept for every transit of a cell by every particle. The tally for density in a particular cell consists of adding the time interval Δt_i during which a particle is in it, every time the cell is crossed. The tally for flux density adds up the time interval multiplied by speed: $v_i \Delta t_i$. After launching and tracking to extinction a large number N of random emitted particles, the sums are

$$S_n = \sum_i \Delta t_i, \qquad S_\phi = \sum_i v_i \Delta t_i.$$

Deduce quantitative physical formulas for what the particle and flux densities are (not just that they are proportional to these sums), and the uncertainties in them, giving careful rationales.

A total of N particles tracked from the source (not including the other particles arising from collisions that we also had to track) is the number of particles that would be emitted in a time $T = N/R$. Suppose the typical physical duration of a particle "flight" (the time from launch to absorption/loss of the particle and all of its descendants) is τ.

If $T \gg \tau$, then it is clear that the calculation is equivalent to treating a physical duration T. In that case almost all particle flights are completed within the same physical duration T. Only those that start within τ of the end of T would be physically uncompleted; and only those that started a time less than τ before the start of T would be still in flight when T starts. The fraction affected is $\sim \tau/T$, which is small. But in fact, even if T is not much longer than τ, the calculation is still equivalent to simulating a duration T. In an actual duration T there would then be many flights that are unfinished at the end of T, and others that are part way through at its start. But on average the different types of partial flights add up to represent a number of total flights equal to N. The fact that in the physical case the partial flights are of different particles, whereas in the Monte Carlo calculation they are of the same particle, does not affect the average tally.

Given, then, that the calculation is effectively simulating a duration T, the physical formulas for particle and flux densities in a cell of volume V are

$$n = S_n/TV = S_n R/NV, \quad \text{and} \quad \phi = S_\phi/TV = S_\phi R/NV.$$

To obtain the uncertainty, we require additional sample sums. The total number of tallies in the cell of interest we'll call $S_1 = \sum_i 1$. We may also need the sum of the squared tally contributions $S_{n^2} = \sum_i \Delta t_i^2$ and $S_{\phi^2} = \sum_i (v_i \Delta t_i)^2$. Then finding S_n and S_ϕ can be considered to be a process of making S_1 random selections of typical transits of the cell from probability distributions whose mean contribution per transit are S_n/S_1 and S_ϕ/S_1, respectively. Of course, the probability distributions aren't actually known, they are indirectly represented by the Monte Carlo calculation. But we don't need to know what the distributions are, because we can invoke the fact that the variance of the mean of a large number (S_1) of samples from a population of variance σ^2 is just σ^2/S_1. We need an estimate of the population variance for density and flux density. That estimate is provided by the standard expressions for variance:

$$\sigma_n^2 = \frac{1}{S_1 - 1}[S_{n^2} - (S_n/S_1)^2], \quad \sigma_\phi^2 = \frac{1}{S_1 - 1}[S_{\phi^2} - (S_\phi/S_1)^2].$$

So the uncertainties in the tally sums when a fixed number $S_1 \approx S_1 - 1$ of tallies occurs are

$$\sigma_n/\sqrt{S_1}, \quad \text{and} \quad \sigma_\phi/\sqrt{S_1}.$$

However, there is also uncertainty arising from the fact that the number of tallies S_1 is *not* exactly fixed, it varies from case to case in different Monte Carlo trials, each of N flights. Generally, S_1 is Poisson distributed, so its variance is equal to its mean $\sigma_{S_1}^2 = S_1$. Variation δS_1 in S_1 gives rise to a variation approximately $(S_n/S_1)\delta S_1$ in S_n. Also, it is reasonable to suppose that the variances of contribution size, σ_n^2 and σ_ϕ^2, are not correlated with the variance arising from the number of samples $(S_n/S_1)^2\sigma_{S_1}^2$, so we can simply add them and arrive at total uncertainty in S_n and S_ϕ (which we write δS_n and δS_ϕ):

$$\delta S_n = \sqrt{\frac{\sigma_n^2 + S_n^2}{S_1}} \qquad \delta S_\phi = \sqrt{\frac{\sigma_\phi^2 + S_\phi^2}{S_1}}$$

Usually, $\sigma_n \lesssim S_n$ and $\sigma_\phi \lesssim S_\phi$, in which case we can (within a factor of $\sqrt{2}$) ignore the σ_n and σ_ϕ contributions, and approximate the *fractional* uncertainty in both density and flux as $1/\sqrt{S_1}$. In this approximation, the squared sums S_{n^2} and S_{ϕ^2} are unneeded.

Exercise 12. Monte Carlo statistics

1. Suppose in a Monte Carlo transport problem there are N_j different types of collision, each of which is equally likely. Approximately what is the

statistical uncertainty in the estimate of the rate of collisions of a specific type i when a total $N_t \gg N_j$ of collisions has occurred:

(a) if for each collision just one contribution to a specific collision type is tallied;

(b) if for each collision a proportionate $(1/N)$ contribution to every collision type tally is made?

If adding a tally contribution for any individual collision type has a computational cost that is a multiple f times the rest of the cost of the simulation:

(c) how large can f be before it becomes counterproductive to add proportions to all collision type tallies for each collision?

2. Consider a model transport problem represented by two particle ranges, low and high energy: 1, 2. Suppose on average there are n_1 and n_2 particles in each range and $n_1 + n_2 = n$ is fixed. The particles in these ranges react with the material background at overall average rates r_1 and r_2.

   (a) A Monte Carlo determination of the reaction rate is based on a random draw for each particle determining whether or not it has reacted during a time T (chosen such that $r_1T, r_2T \ll 1$). Estimate the fractional statistical uncertainty in the reaction-rate determination after drawing n_1 and n_2 times, respectively.

   (b) Now consider a determination using the same $r_{1,2}$, T, and total number of particles, n, but distributed differently so that the numbers of particles (and hence number of random draws) in the two ranges are artificially adjusted to n'_1, n'_2 (keeping $n'_1 + n'_2 = n$), and the reaction-rate contributions are appropriately scaled by n_1/n'_1 and n_2/n'_2. What is now the fractional statistical uncertainty in reaction-rate determination? What is the optimum value of n'_1 (and n'_2) to minimize the uncertainty?

3. Build a program to sample randomly from an exponential probability distribution $p(x) = \exp(-x)$, using a built-in or library uniform random deviate routine. Code the ability to form a series of M independent samples labelled j, each sample consisting of N independent random values x_i from the distribution $p(x)$. The samples are to be assigned to K bins depending on the sample mean $\mu_j = \sum_{i=1}^{N} x_i/N$. Bin k contains all samples for which $x_{k-1} \leq \mu_j < x_k$, where $x_k = k\Delta$. Together they form a distribution n_k, for $k = 1, \ldots, K$, that is the number of samples with μ_j in the bin k. Find the mean $\mu_n = \sum_{k=1}^{K} n_k(k - 1/2)\Delta/M$ and variance $\sigma_n^2 = \sum_{k=1}^{K} n_k[(k - 1/2)\Delta - \mu_n]^2/(M - 1)$ of this distribution n_k, and compare them with the prediction of the Central Limit Theorem, for $N = 100$, $K = 30$, $\Delta = 0.2$,

and two cases: $M = 100$ and $M = 1000$. Submit the following as your solution:

1. Your code in a computer format that is capable of being executed.
2. Plots of n_k versus x_k, for the two cases.
3. Your numerical comparison of the mean and variance, and comments as to whether it is consistent with theoretical expectations.

13

Next steps

This book is by deliberate choice an introduction to numerical methods for scientists and engineers that is *concise*. The idea is that brevity is the best way to grasp the big picture of computational approaches to problem solving. However, undoubtedly for some people the conciseness overstrains your background knowledge and requires an uncomfortably accelerated learning curve. If so, you may benefit by supplementing your reading with a more elementary text.[1]

For readers who have survived thus far without much extra help, congratulations! If you've really made the material your own by doing the exercises, you have a wide-ranging essential understanding of the application of numerical methods to physical science and engineering. That knowledge includes some background derivations and some practical applications, and will serve you in good stead in a professional career. It might be all you need, but it is by no means comprehensive

The conciseness has been achieved at the expense of omitting some topics that are without question important in certain applications. It is the purpose of this concluding chapter to give pointers, even more abbreviated than the preceding text, to some of these topics, and so open the doors for students who want to go further. Of course, all of this chapter is enrichment, and demands somewhat deeper thought in places. If for any topic you don't "get it" on a first reading, then don't be discouraged. References to detailed advanced textbooks are given.

[1] One such book with lots of engineering examples is S. C. Chapra and R. P. Canale (2006), *Numerical Methods for Engineers*, fifth edition, or later, McGraw-Hill, New York.

13.1 Finite-element methods

We have so far omitted two increasingly important approaches to solving problems for complicated boundary geometries: unstructured meshes and finite elements. Unstructured meshes enable us to accommodate in a natural way boundaries that are virtually as complicated as we like. The reason they are generally linked with finite-element techniques is that the finite-elements approach offers a systematic way to discretize partial differential equations on unstructured meshes. By contrast, it is far more ambiguous how to implement consistently finite differences on an unstructured mesh. Finite elements are somewhat less intuitive than finite differences, and somewhat more complex. On structured meshes they offer little advantage to compensate; and often result in difference schemes mathematically equivalent to finite differences. So there is far less incentive to use finite elements on structured meshes.

The crucial difference with finite elements lies in the way the approximation to the differential equation is formulated. We've seen that many of the equations we need to solve in physics and engineering are *conservation equations*. They can be expressed in differential form or in integral form. Most often they are derived in integral form, and then, recognizing that the domain over which the conservation applies is arbitrary, we conclude that the integrand must itself be zero. That is the differential form. The finite-elements approach expresses the problem as being to minimize the weighted integral of a finite representation of the equation. So it is in a sense a return to integral form, but with a specific set of weightings that we will explain in a moment.

Consider an elliptic equation of the type that arises from diffusion and many other conservation principles:

$$\nabla.(D\nabla\psi) + s(\boldsymbol{x}) = 0. \tag{13.1}$$

Suppose we multiply this equation by a weight function $w(\boldsymbol{x})$, and recognize that (for a differentiable w)

$$w\nabla.(D\nabla\psi) = \nabla.(wD\nabla\psi) - D(\nabla w).(\nabla\psi). \tag{13.2}$$

When we integrate it over the entire solution domain volume V whose surface is ∂V, then using Gauss's (divergence) theorem, we get

$$\begin{aligned} 0 &= \int_V [w\nabla.(D\nabla\psi) + ws]d^3x \\ &= \int_V [-D(\nabla w).(\nabla\psi) + ws]d^3x + \int_{\partial V} wD\nabla\psi.\boldsymbol{dS}. \end{aligned} \tag{13.3}$$

Remember that, provided ψ exactly satisfies the original differential equation, this integral equation is exactly satisfied for all possible weight functions, w. This fact is expressed by saying that the integral is a "weak form" of the differential equation. However, we are going to represent ψ by a functional form that has only a discrete number of parameters. Generically,

$$\psi(\boldsymbol{x}) = \phi_b + \sum_{k=1}^{N} a_k \psi_k(\boldsymbol{x}), \tag{13.4}$$

where ϕ_b is a known function that satisfies the boundary conditions, ψ_k is a discrete set of functions, and a_k are the parameters consisting of a set of coefficients to be found. This discrete ψ-representation will therefore satisfy the differential equation only approximately.

Boundary conditions on the surface ∂V are important. To avoid getting bogged down in discussing them, we assume that Dirichlet (fixed ψ) conditions are applied. By expressing ψ as the sum of the part we are solving for, that satisfies *homogeneous* boundary conditions $\psi = 0$, plus some known function ϕ_b that satisfies the inhomogeneous conditions but not the original equation, we simplify the functions ψ_k. They are all zero on the boundary.

One benefit of the final right-hand side of eq. (13.3) is that we now have only first-order derivatives of the dependent variable ψ. We can therefore permit ourselves the freedom of a ψ-representation with discontinuous gradient without inducing the infinities that would occur if we used the second-order form $\nabla.(D\nabla\psi)$. We still usually require ψ to be everywhere continuous to avoid infinities from $\nabla\psi$. The general approach of finite elements is to require the *residual*, consisting of the right-hand side of eq. (13.3), to be as close as possible to zero by adjusting the parameters determining ψ. That optimized situation will be the solution.

Of course, we need at least N equations to determine all the coefficients a_k. These will be obtained from different choices of the weight function w. Naturally we also represent the range of possible weight functions discretely with a limited number of parameters (but at least N). The most common choice of w representation is to take it to be represented by exactly the same functions ψ_k. This choice makes sense since it allows the weight function approximately the same freedom as ψ itself; it would be unprofitable to apply far more detailed and flexible w functions than can be represented by ψ. This choice, which goes by the name *Galerkin method*, also gives rise to symmetric matrices, which is often advantageous.

Substituting the total ψ representation of eq. (13.4) in the integral expression eq. (13.3), and using for w, one after the other, the functions ψ_k, we get a set of N equations which can be written in matrix form as

$$\mathbf{Ka} = \mathbf{f}, \tag{13.5}$$

where $\mathbf{a}$ is the column vector of coefficients a_k to be determined. The $N \times N$ matrix $\mathbf{K}$ is symmetric, with coefficients

$$K_{jk} = \int_V \nabla\psi_j.\nabla\psi_k D \, d^3x. \tag{13.6}$$

And the column vector $\mathbf{f}$ has coefficient values

$$f_j = \int_V [\psi_j s - \nabla\psi_j.\nabla\phi_b D] \, d^3x. \tag{13.7}$$

In the mechanics field, which was where finite-element techniques mostly grew up, $\mathbf{K}$ is called the stiffness matrix, $\mathbf{f}$ the force vector, and $\mathbf{a}$ the displacement vector.

This treatment has shown in principle how an integral approach can reduce our partial differential equation to a matrix equation, to which the various approaches to matrix inversion can be applied to find the solution. But to be specific, we need to decide what to use for the basis functions ψ_k. They are each defined over the entire domain, but it will greatly reduce the number of non-zero entries of $\mathbf{K}$ if we choose basis functions that are localized, i.e. they have non-zero value only in a small region of the domain. Then when j and k refer to functions that are localized in *non-overlapping* local regions, their corresponding overlap integral K_{jk} is zero. Various choices of localized basis function are possible, but if we think of the function as being defined at a mesh of *nodes* $\boldsymbol{x}_k$, then a piecewise linear representation arises from basis functions that are identified with each node and vary linearly from unity at the node to zero at the adjacent nodes. In one dimension, these are "triangle functions" (sometimes called "tent" functions), whose derivatives are bipolar "box functions" as illustrated in Fig. 13.1. Such a set of functions will give rise to a matrix $\mathbf{K}$ that is tridiagonal, just as was the case for finite differences in one dimension. Using smoother, higher-order, functions representing the elements is sometimes advantageous. Cubic Hermite functions, and cubic B-splines, are two examples. They lead to a matrix that is not quite as sparse, possessing typically seven non-zero diagonals (in one dimension). Their ability to treat higher-order differential equations efficiently compensates for the extra computational complexity and cost.

In multiple dimensions, the geometry becomes somewhat more complicated, but, for example, there is a straightforward extension of the piecewise linear treatment to an unstructured mesh of tetrahedra in three dimensions. For any point within a particular tetrahedron, the interpolated value of the function

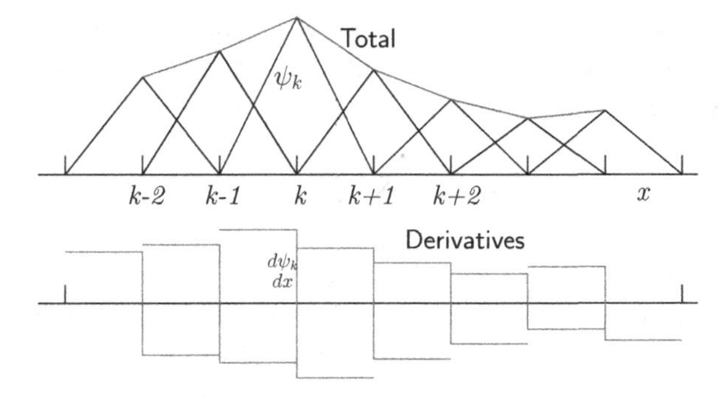

Figure 13.1 Localized triangle functions in one dimension, multiplied by coefficients, sum to a piecewise linear total function. Their derivatives are box functions with positive and negative parts. They overlap only with adjacent functions.

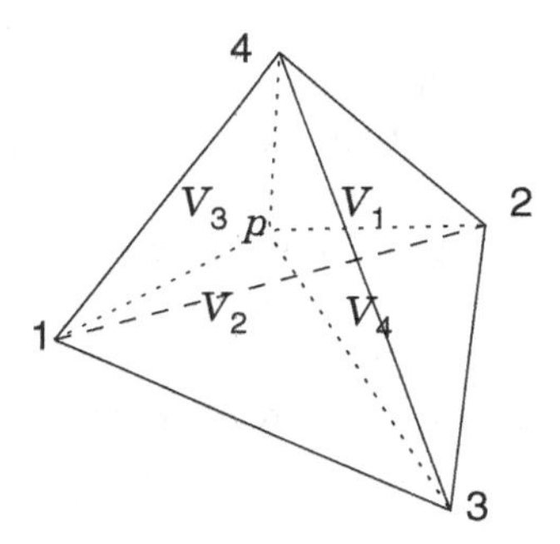

Figure 13.2 In linear interpolation within a tetrahedron, the lines from a point p to the corner nodes of the tetrahedron, divide it into four smaller tetrahedra, whose volumes sum to the total volume. The interpolation weight of node k is proportional to the corresponding volume, V_k.

is taken to be equal to a weighted sum of the values at the four corner nodes. The weight of each corner is equal to the volume of the smaller tetrahedron obtained by replacing the corner with the point of interest, divided by the total volume of the original tetrahedron, which Fig. 13.2 illustrates.[2] The ψ_k

[2] The volume of a tetrahedron can be written as a 4×4 determinant using the cartesian coordinates (x_k, y_k, z_k) of its four corners ($k = 1, 4$) as

$$V = \frac{1}{6}\begin{vmatrix} 1 & x_1 & y_1 & z_1 \\ 1 & x_2 & y_2 & z_2 \\ 1 & x_3 & y_3 & z_3 \\ 1 & x_4 & y_4 & z_4 \end{vmatrix}.$$

function associated with a node k is unity at that node and decreases linearly to zero along every connection leg to an adjacent node. The interpolation formula within a single tetrahedral element is then

$$\psi(p) = \sum_1^4 V_k a_k / V. \tag{13.8}$$

In the resulting matrix **K** the only non-zero overlap integrals (coefficients) for each row j are those for nodes k that connect to node j. The number of such connections is generally small, amounting typically to about ten per node, but depending on the details of the mesh. Therefore, **K** is very sparse; and the necessary computational effort to obtain the solution can be greatly reduced when advanced sparse-matrix techniques are employed.

More complicated meshes are also possible, using as elements other polyhedra such as hexahedra (non-rectangular cubes). Also higher-order interpolations of the solution within the elements are sometimes used. These lead to increasing complications in handling the geometrical aspects of the problem. (Even without these enhancements, the coding complications are already significant.) Generally, mesh-generation libraries or applications are used to construct the nodal mesh and its connections. Then the overlap integrals need to be evaluated to construct the matrix **K** and the vector **f**.[3] In addition to open-source libraries, many commercial computing packages provide the mechanisms for mesh generation, integration, and equation discretization, often with substantial graphical user interfaces to ease the process.

13.2 Discrete Fourier analysis and spectral methods

A function $f(t)$, defined over a finite range of its argument $0 \le t < T$, can be expressed in terms of a Fourier series of discrete frequencies $\omega_n = 2\pi\nu_n = 2\pi n/T$ as

$$f(t) = \sum_{n=-\infty}^{\infty} e^{i\omega_n t} F_n, \tag{13.9}$$

[3] These important details are discussed in texts devoted to finite elements, such as, T. J. R. Hughes (1987), *The Finite Element Method*, Prentice Hall, Englewood Cliffs, NJ. The classic text on the background mathematical theory is G. Strang and G. J. Fix (1973, 2008), *An Analysis of the Finite Element Method*, Reissued by Wellesley-Cambridge, Wellesley, Press, MA.

where the Fourier coefficient for integer n is

$$F_n = \frac{1}{T}\int_0^T e^{-i\omega_n t} f(t)dt. \qquad (13.10)$$

Actually the expression (13.9) gives a function $f(t)$ that is *periodic* in t with period T.

Numerical representations are not usually continuous; they are discrete. So suppose the function f is given only at uniformly spaced argument values $t_j = j\Delta t$ for $j = 0, 1, \ldots, N-1$, where $\Delta t = T/N$, and write $f(t_j) = f_j$. Then these N discrete values[4] of f allow us to determine only N Fourier coefficients. It is intuitive that these should be the N *lowest* (absolute) frequency components, since one clearly cannot meaningfully represent frequencies higher than roughly $1/\Delta t$ on this discrete mesh. The lowest N frequencies are those for which $-N/2 < n \le N/2$. This observation immediately refines the qualification of how high a frequency can be represented on a discrete mesh. The highest frequency[5] has $|n| = N/2$, giving $\omega_n = 2\pi\nu_n = 2\pi N/2T = \pi/\Delta t$. This limit is called the Nyquist limit.

If, discarding all pretence at rigor, we decide to approximate f as a sum of Dirac delta functions $f(t) \approx \sum_j \Delta t f_j \delta(t - t_j)$ (which gives approximately correct integrals), then the integral for F_n (13.10) can be expressed approximately as a sum:[6]

$$F_n = \frac{1}{T}\int_{0-\epsilon}^{T-\epsilon} e^{-i\omega_n t} f(t)dt \approx \frac{1}{T}\sum_{j=0}^{N-1} e^{-i\omega_n t_j} f_j \Delta t = \frac{1}{N}\sum_{j=0}^{N-1} e^{-i2\pi nj/N} f_j. \qquad (13.11)$$

Now we notice that the final expression for F_n is *periodic* with period N; that is, $F_{n+N} = F_n$. It is therefore convenient to consider the range of frequency index $-N/2 < n \le N/2$ to be re-ordered by replacing every negative n by the positive value $n + N$. Then n runs from 0 to $N - 1$. One should remember, though, that the upper half of this range really represents negative frequencies. Incidentally, the periodicity of F_n illustrates that one can equally well consider F_n to represent *any* frequency $\nu = (n + kN)/T$ with k an integer. That is because sampling a continuous function $e^{i2\pi(n+kN)t/T}$ at $t = j\Delta t = jT/N$ gives values independent of k. Thus, all frequencies higher than the Nyquist limit $\nu = N/2T$, contained in a continuous signal sampled discretely, are shifted

[4] Since f is periodic, its value at $j = 0$ is the same as at $j = N$, so the range is equivalent to $1, \ldots, N$.

[5] If N is odd, then the highest frequency is obviously $(N - 1)/2$, but we'll avoid reiterating this alternative.

[6] ϵ is a small adjustment of the end point to avoid it falling on a delta function.

into the Nyquist range and appear at a lower absolute frequency in the sampled representation. This effect is called aliasing.

The ***discrete* Fourier transform** and its inverse can thus be considered to be

$$f_j = \sum_{n=0}^{N-1} e^{i2\pi nj/N} F_n \; ; \qquad F_n = \frac{1}{N} \sum_{j=0}^{N-1} e^{-i2\pi nj/N} f_j. \tag{13.12}$$

Substituting for F_n in this expression for f_j, one obtains coefficients that are sums of the form $\sum_n e^{i2\pi nj/N} e^{-i2\pi nj'/N}$, which is zero if $j \neq j'$ and N if $j = j'$. Therefore, these two equations are exact, and could have been simply adopted as definitions from the start. The discrete Fourier transforms need not be considered *approximations* to continuous transforms; but it is helpful to recognize their relationship to Fourier series representations of continuous functions.

Of course we aren't really interested in quantities that consist of a set of delta functions. If, instead of simply multiplying by Δt, we convolve $\sum_j f_j \delta(t - t_j)$ with a triangle function that is unity at $t = 0$ and descends linearly to zero at $t = \pm\Delta t$, then we will recover a piecewise linear function whose values at t_j are f_j. This is a much closer representation of the sort of function we might be considering.

The Fourier transform of a convolution of two functions is the product of their Fourier transforms.

$$\int_0^T e^{-\omega_n t} \int_0^T f(t')g(t-t')dt'dt = \int_0^T e^{-\omega_n t'} f(t')dt' \int_0^T e^{-\omega_n t''} g(t'')dt''. \tag{13.13}$$

Therefore, the Fourier coefficients F_n corresponding to the piecewise linear form of f are the expressions in eq. (13.12) multiplied by the scaled Fourier transform of the triangle function, namely

$$K_n = \int_{-\Delta t}^{\Delta t} e^{-i\omega_n t}(1 - |t|/\Delta t)dt/\Delta t = \frac{\sin^2(\omega_n \Delta t/2)}{(\omega_n \Delta t/2)^2}. \tag{13.14}$$

So when dealing with continuous functions, and interpolating using the inverse transform, eq. (13.9), one ought probably to filter the signal in frequency space. Multiplying by the sinc-squared function $K_n = \sin^2(z_n)/z_n^2$, with $z_n = \omega_n \Delta t/2$, is the equivalent of doing piecewise linear interpolation. At the highest absolute frequency, $\omega_{N/2}$, where $\omega_n \Delta t/2 = \pi N \Delta T/T = \pi$, the filter reaches its first zero. It modestly limits the frequency bandwidth of the signal.

It turns out to be possible to evaluate discrete Fourier transforms (by successively dividing the domain in half) much faster than implementing the $N \times N$ multiplications implied by performing the sums in eq. (13.12) one after the other. The algorithms that reduce the computational effort to of order $N \ln N$ multiplications are called fast Fourier transforms.[7] They, naturally enough, find important uses in spectral analysis and filtering. But, less obviously, they also provide powerful techniques for solving partial differential equations by representing the solution in terms of a finite number of Fourier coeffients, as an alternative to a finite number of discrete values.

Spectral representation is most powerful for linear problems in which there is an ignorable coordinate because of some inherent symmetry. In such situations each Fourier component becomes independent of the others. What's more, the Fourier components are the eigenmodes (in the direction of symmetry) of the problem. Consequently, it is sometimes the case that including a very few such Fourier components, even as few as one, can represent the solution. In effect, this practically lowers the dimensionality of the problem by one, leading to major computational advantage. The equations can be formulated in terms of Fourier modes in the symmetry direction and finite differences in the other direction(s). For separable linear problems like this, there is actually no pressing need for the Fourier transforms to be *fast*, because the solution in space only needs to be reconstructed after the solution in terms of Fourier modes has been found.

If the equations to be solved are non-linear, however, or the symmetry is only approximate, then different Fourier modes are coupled together. Then a larger number of modes is necessary, and the computational advantage becomes less. Even so, there are sometimes substantial remaining benefits to a spectral representation, provided the Fourier transform is fast. The advantages can be understood as follows.

Suppose we have a partial differential equation in which we wish to represent the x coordinate by a Fourier expansion. The representations of the other coordinates and their differentials do not affect this question. Consider the equation to consist of a part that is linear in the dependent variable u, $\mathcal{L}(u)$, and a part that is quadratic, $\mathcal{M}(u)\mathcal{N}(u)$; so

$$\mathcal{L}(u) + \mathcal{M}(u)\mathcal{N}(u) = s(x). \tag{13.15}$$

Here $\mathcal{L}$, $\mathcal{M}$, $\mathcal{N}$ are linear x-differential operators, for example: $\mathcal{L}(u) = \frac{\partial u}{\partial x} + gu$. We represent u by a sum of N Fourier modes $u = \sum_{n=1}^{N} \mathrm{e}^{ik_n x} U_n$, where for an x-domain of length X, $k_n = 2\pi n/X$. Substituting a Fourier mode into a linear operator gives rise to an algebraic multiplier. For example $\left(\frac{\partial}{\partial x} + g\right) \mathrm{e}^{ik_n x} = (ik + g)\mathrm{e}^{ik_n x}$. So $\mathcal{L}(u) = \sum_n L_n \mathrm{e}^{ik_n x}$, where L_n is the multiplier arising from

[7] An extensive discussion is given in *Numerical Recipes*, chapter 12. Various open-source implementations are available.

$\mathcal{L}$ for the k_n mode. The equations that determine the U_n are found by Fourier transforming the differential equation we started with:

$$\frac{1}{X}\int_0^X [\mathcal{L}(u) + \mathcal{M}(u)\mathcal{N}(u)]\mathrm{e}^{-ik_m x}dx = S_m. \qquad (13.16)$$

Substituting in the Fourier expansion of u, and using the orthonormal properties of the modes: $\int_0^X \mathrm{e}^{ik_n x}\mathrm{e}^{-ik_m x}dx/X = \delta_{mn}$ we get

$$L_m U_m \mathrm{e}^{ik_m x} + \sum_{l+n=m} M_l U_l \mathrm{e}^{ik_l x} N_n U_n \mathrm{e}^{ik_n x} = S_m \qquad (13.17)$$

The sum arising from the quadratic term couples all the individual modal equations together. Without it they would be uncoupled. The coupling term has the form of a convolution sum. To evaluate the sum directly requires for each equation m that we evaluate, on average, $N/2$ terms, for a total computational multiplication count of approximately $2N^2$ (because there are four multiplications per term) per solution step. Generally, a non-linear equation must be solved by iterative steps in which the non-linear term must be evaluated. An alternative way to evaluate the non-linear term at each step is to transform back to x-space, perform the multiplication on the x-space $\mathcal{M}(u)\mathcal{N}(u)$, and then FFT the product again to obtain the sum term for the Fourier mode equation (13.17). The differencing to evaluate the operators $\mathcal{M}$ and $\mathcal{N}$ costs a few multiplications, say p. The product for all N values of U will then cost Np. And the two FFTs will be only approximately $2N \ln N$. Therefore the total cost of this alternative scales like $N(2\ln N + p)$. Thus, using the FFT approach reduces the $2N^2$ computational cost scaling to $N(2\ln N + p)$.

Some extra questions arise concerning aliasing, but this FFT approach is nevertheless used to good advantage for some applications.

13.3 Sparse-matrix iterative Krylov solution

We stopped our development of iterative linear system solution, which we saw in Chapter 6 is at the heart of solving most boundary-value differential equations, before introducing the most significant modern developments in this field. These are associated with the name of Krylov. Here we give the barest introduction. A modern textbook should be consulted for more details. First, as far as terminology is concerned, the name refers to the use of the *subspace* of a vector space that is accessed by repeated multiplications by the same matrix. The Krylov subspace $\mathcal{K}_k(\mathbf{A}, \mathbf{b})$ of dimension k generated by a matrix $\mathbf{A}$ from an initial vector **b** consists of all vectors that can be expressed in the form of a sum of coefficients times the vectors $\mathbf{A}^j\mathbf{b}$, for $j = 0, 1, \ldots, k-1$. The expression $\mathbf{A}^j$ means *matrix* multiply the unit matrix **I** by **A**, j times.

The reason why this subspace is so important is that it encompasses all the vectors that can be accessed by an iterative scheme that uses simply addition, scaling, and multiplication by **A**. Multiplication by a sparse matrix involves far fewer computations than by a full matrix, or than inversion of a sparse matrix. Therefore, iterative solution techniques that require only sparse matrix *multiplications* will be fast. Indeed the matrix itself may never need to be formed, all that is needed is an algorithm for multiplying by it. Let's consider how we might construct such an interative scheme for solving

$$\mathbf{A}\mathbf{x} = \mathbf{b} \tag{13.18}$$

for **x** given **b**. After say k iterations, we have a vector $\mathbf{x}_k$, which we hope is nearly the solution. The extent to which it is not yet the solution is given by the extent to which the residual $\mathbf{r}_k \equiv (\mathbf{b} - \mathbf{A}\mathbf{x}_k)$ is non-zero. If the matrix **A** is not very different from the identity matrix, then an intuitive scheme for obtaining the next iteration vector $\mathbf{x}_{k+1}$ would be to add the residual to $\mathbf{x}_k$ and take $\mathbf{x}_{k+1} = \mathbf{x}_k + \mathbf{r}_k$. In practice it is better to use an increment $\mathbf{p}_k$ chosen to be like $\mathbf{r}_k$, but "conjugate" to previous increments (in a sense to be explained), and scaled by a coefficient α_k chosen to minimize the resulting residual $\mathbf{r}_{k+1}$. Thus $\mathbf{x}_{k+1} = \mathbf{x}_k + \alpha_k\mathbf{p}_k$, which means

$$\mathbf{r}_{k+1} = \mathbf{r}_k - \alpha_k\mathbf{A}\mathbf{p}_k. \tag{13.19}$$

Different schemes use different choices of $\mathbf{p}_k$ and α_k. The search direction $\mathbf{p}_k$, is chosen to be

$$\mathbf{p}_k = \mathbf{P}\mathbf{r}_k + \sum_{j=0}^{k-1} \beta_{kj}\mathbf{p}_j \tag{13.20}$$

where **P** is an optional matrix, omitted (or, equivalently, the identity) in the simplest case. The coefficients β_{kj} are chosen so as to satisfy the conjugacy condition that we'll specify in a moment (eq. 13.23).

If we start from an initial vector $\mathbf{x}_0 = \mathbf{0}$, so $\mathbf{r}_0 = \mathbf{b}$ (without loss of generality[8]), each $\mathbf{P}\mathbf{r}_k$ and $\mathbf{p}_k$ generated by the iteration consists of sums of terms like $(\mathbf{PA})^j\mathbf{P}\mathbf{b}$ with $0 \leq j < k$. In other words, they are members of

[8] A non-zero $\mathbf{x}_0$ can be removed by reexpressing the equation in terms of $\mathbf{x}' = \mathbf{x} - \mathbf{x}_0$.

the Krylov subspace $\mathcal{K}_k(\mathbf{PA}, \mathbf{Pb})$. The general description "Krylov" technique applies to all approaches that use this repeated multiplication process.[9]

We suppose that minimizing the residual is taken to mean minimimizing a bilinear form $\mathbf{r}_{k+1}^T \mathbf{R} \mathbf{r}_{k+1}$, where $\mathbf{R}$ is a fixed symmetric matrix to be chosen later. Setting the differential of this form with respect to α_k equal to zero we get

$$(\mathbf{A}\mathbf{p}_k)^T \mathbf{R}(\mathbf{r}_k - \alpha_k \mathbf{A}\mathbf{p}_k) = \mathbf{p}_k^T \mathbf{A}^T \mathbf{R} \mathbf{r}_{k+1} = 0, \qquad (13.21)$$

whose solution is

$$\alpha_k = \mathbf{p}_k^T \mathbf{A}^T \mathbf{R} \mathbf{r}_k / \mathbf{p}_k^T \mathbf{A}^T \mathbf{R} \mathbf{A} \mathbf{p}_k, \qquad (13.22)$$

and which gives a new residual $\mathbf{r}_{k+1}$ that is orthogonal (in the sense of eq. (13.21): $\mathbf{p}_k^T \mathbf{A}^T \mathbf{R} \mathbf{r}_{k+1} = 0$) to the search direction. The conjugacy condition on the search directions is then taken to be that

$$\mathbf{p}_j^T \mathbf{A}^T \mathbf{R} \mathbf{A} \mathbf{p}_k = 0 \quad \text{for} \quad j \neq k, \qquad (13.23)$$

which requires[10] (substituting from eq. (13.20))

$$\beta_{kj} = -\mathbf{p}_j^T \mathbf{A}^T \mathbf{R} \mathbf{A} \mathbf{P} \mathbf{r}_k / \mathbf{p}_j^T \mathbf{A}^T \mathbf{R} \mathbf{A} \mathbf{p}_j. \qquad (13.24)$$

The residual minimization process produces a series of residuals and search directions having orthogonality properties that can be used to advantage. The first property is *mutual orthogonality*, which is that

$$\mathbf{p}_j^T \mathbf{A}^T \mathbf{R} \mathbf{r}_k = 0 \quad \text{for all } k \text{ and all } j < k. \qquad (13.25)$$

This fact follows inductively, assuming that the condition holds up to k. Premultiplying eq. (13.19) by $\mathbf{p}_j \mathbf{A}^T \mathbf{R}$ gives

$$\mathbf{p}_j^T \mathbf{A}^T \mathbf{R} \mathbf{r}_{k+1} = \mathbf{p}_j \mathbf{A}^T \mathbf{R} \mathbf{r}_k - \alpha_k \mathbf{p}_j \mathbf{A}^T \mathbf{R} \mathbf{A} \mathbf{p}_k. \qquad (13.26)$$

For $j < k$ the right-hand side's first term is zero by hypothesis and the second is zero by conjugacy (13.23). For $j = k$, the left-hand side is zero by eq. (13.21). Therefore mutual orthogonality holds up to $k + 1$ and so for all k.

The second property is *residual orthogonality*. This follows from mutual orthogonality. Because $\mathbf{p}_j$ and $\mathbf{P}\mathbf{r}_j$ ($j < k$) span the same Krylov subspace,

[9] The present derivation draws on C. P. Jackson and P. C. Robinson (1985) A numerical study of various algorithms related to the preconditioned conjugate gradient method, *International Journal for Numerical Methods in Engineering*, **21**, 1315-1338, and G. Markham (1990), Conjugate gradient type methods for indefinite, asymmetric, and complex systems, *IMA Journal of Numerical Analysis*, **10**, 155–170.

[10] This is a type of Gramm–Schmidt basis-set construction.

orthogonality of a vector to all of one set implies orthogonality to all of the other. Therefore we can replace the $\mathbf{p}_j$ in eq. (13.25) with $\mathbf{Pr}_j$ and conclude

$$(\mathbf{Pr}_j)^T\mathbf{A}^T\mathbf{Rr}_k = \mathbf{r}_k^T\mathbf{R}^T\mathbf{APr}_j = 0 \quad \text{for all } k \text{ and all } j < k. \tag{13.27}$$

The final property is obtained by premultiplying eq. (13.20) by $\mathbf{r}_k^T\mathbf{R}^T\mathbf{A}$, which demonstrates that the non-zero scalar products with $\mathbf{r}_k$ of $\mathbf{p}_k$ and $\mathbf{Pr}_k$ are the same:

$$\mathbf{r}_k^T\mathbf{R}^T\mathbf{Ap}_k = \mathbf{r}_k^T\mathbf{R}^T\mathbf{APr}_k. \tag{13.28}$$

The first of these is equal to the denominator of α_k in eq. (13.22).

Residual orthogonality enables us to demonstrate conditions on an expression nearly the same as the numerator of β_{kj}. We premultiply a version of eq. (13.19) using index j by $\mathbf{r}_k^T\mathbf{R}^T\mathbf{AP}$ to get (for $j < k$)

$$\mathbf{r}_k^T\mathbf{R}^T\mathbf{APr}_{j+1} = \mathbf{r}_k^T\mathbf{R}^T\mathbf{APr}_j - \alpha_j\mathbf{r}_k^T\mathbf{R}^T\mathbf{APAp}_j. \tag{13.29}$$

The first term on the right-hand side is zero by residual orthogonality. The left-hand side is also zero by residual orthogonality except when $j = k - 1$. Therefore, the combination $\mathbf{r}_k^T\mathbf{R}^T\mathbf{APAp}_j = \mathbf{p}_j^T\mathbf{A}^T\mathbf{P}^T\mathbf{A}^T\mathbf{Rr}_k$ is zero except when $j = k - 1$. This last form is equal to the numerator of β_{kj} provided that

(1) $\mathbf{P}$ and $\mathbf{A}$ are symmetric,
(2) $(\mathbf{AP})$ and $(\mathbf{AR})$ commute.

Those conditions are sufficient to ensure that only one of the β_{kj}, namely $\beta_{k,k-1}$, is non-zero. It can be rewritten

$$\beta_{k,k-1} = \mathbf{r}_k^T\mathbf{R}^T\mathbf{APr}_k/\mathbf{r}_{k-1}^T\mathbf{R}^T\mathbf{APr}_{k-1}, \tag{13.30}$$

and α_k is also generally rewritten as

$$\alpha_k = \mathbf{r}_k^T\mathbf{R}^T\mathbf{APr}_k/\mathbf{p}_k^T\mathbf{A}^T\mathbf{RAp}_k. \tag{13.31}$$

There is a major advantage to the property of having only one non-zero β coefficient, which we'll call the "currency-property," It is that the iteration requires only the *current* residual and search direction vectors, rather than all the previous vectors back to zero. In big problems, the vectors are long. Keeping all of them would require a costly increase of storage space, and of arithmetic.

Having done the derivation for a general choice of bilinear matrix $\mathbf{R}$ and matrix $\mathbf{P}$, we can consider various different popular iteration schemes[11] as examples with different choices of $\mathbf{R}$ and $\mathbf{P}$.

The **conjugate gradient** algorithm takes $\mathbf{R} = \mathbf{A}^{-1}$ and $\mathbf{P} = \mathbf{I}$. The inverse of $\mathbf{A}$ is not required to be calculated because $\mathbf{R}$ (which equals $\mathbf{R}^T$) always appears multiplied by $\mathbf{A}$. Conjugacy is then $\mathbf{p}_j^T\mathbf{A}\mathbf{p}_k = 0$, and orthogonality is $\mathbf{r}_j^T\mathbf{r}_k = 0$. We require $\mathbf{A}$ to be symmetric for the scheme to work.

A different choice is called the **minimum residual** scheme, $\mathbf{R} = \mathbf{I}$ and $\mathbf{P} = \mathbf{I}$, in which α is chosen to mimimize $\mathbf{r}_{k+1}^T\mathbf{r}_{k+1}$, with the result that conjugacy is $\mathbf{p}_j\mathbf{A}^T\mathbf{A}\mathbf{p}_k$ and mutual orthogonality is $\mathbf{p}^T\mathbf{A}^T\mathbf{r}_k = 0$. This scheme again satisfies the conditions for only $\beta_{k,k-1}$ to be non-zero, provided that $\mathbf{A}$ is symmetric. It then has efficiency similar to the conjugate gradient scheme.

If $\mathbf{A}$ is *not* symmetric, then if we wish the currency property to hold, permitting us to retain only current vectors, we must implement a compound scheme in which the compound matrix $\mathbf{A}_c$ is symmetric. The best way to do this is usually the **bi-conjugate gradient** scheme, in which the iteration is applied to compound vectors $\mathbf{x}_c = \begin{pmatrix}\bar{\mathbf{x}}\\ \mathbf{x}\end{pmatrix}$, $\mathbf{p}_c = \begin{pmatrix}\bar{\mathbf{p}}\\ \mathbf{p}\end{pmatrix}$, and $\mathbf{r}_c = \begin{pmatrix}\mathbf{r}\\ \bar{\mathbf{r}}\end{pmatrix}$ with $\mathbf{A}_c = \begin{pmatrix}\mathbf{0} & \mathbf{A}\\ \mathbf{A}^T & \mathbf{0}\end{pmatrix}$, $\mathbf{R}_c = \mathbf{A}_c^{-1}$ and $\mathbf{P}_c = \begin{pmatrix}\mathbf{0} & \mathbf{I}\\ \mathbf{I} & \mathbf{0}\end{pmatrix}$. Since $\mathbf{A}_c\mathbf{R_c} = \mathbf{I}_c$ the commutation requirements are immediately fulfilled. The extra costs of this scheme compared with the conjugate gradient scheme are its doubled vector length,[12] and the requirement to be able to multiply by $\mathbf{A}^T$ as well as by $\mathbf{A}$. It is generally much more efficient than other approaches to symmetrization, such as writing $\mathbf{A}^T\mathbf{A}\mathbf{x} = \mathbf{A}^T\mathbf{b}$, even when $\mathbf{A}^T\mathbf{A}$ retains sparsity properties. The bi-conjugate gradient scheme has essentially the same range of eigenvalues as the original matrix, whereas $\mathbf{A}^T\mathbf{A}$ squares the eigenvalues.

[11] Succinct descriptions of a wide range of iterative algorithms are given by R. Barrett, M. Berry, T. F. Chan, *et al.* (1994) *Templates for the Solution of Linear Systems: Building Blocks for Iterative Methods*, second edition, SIAM, Philadelphia, which is available at http://www.netlib.org/linalg/html_templates/report.html. Open-source libraries of implementations of Krylov schemes are widely available. One of the more comprehensive is "SLATEC" at http://www.netlib.org/slatec/.

[12] Bi-conjugate gradient iterations are implemented using two residual and two search direction vectors requiring just one multiplication by $\mathbf{A}$ and one by $\mathbf{A}^T$ per step in this way:

1 Initialize:	choose $\mathbf{x}_0, \mathbf{r}_0 = \mathbf{b} - \mathbf{A}\mathbf{x}_0, \mathbf{p}_0 = \mathbf{r}_0$, choose $\bar{\mathbf{r}}_0, \bar{\mathbf{p}}_0 = \bar{\mathbf{r}}_0$, set $k = 1$.
2 Calculate α:	$\alpha_{k-1} = \mathbf{r}_{k-1}^T\bar{\mathbf{r}}_{k-1}/\mathbf{p}_{k-1}^T\mathbf{A}\bar{\mathbf{p}}_{k-1}$
3 Update $\mathbf{r}$s, $\mathbf{x}$:	$\mathbf{r}_k = \mathbf{r}_{k-1} - \alpha_{k-1}\mathbf{A}\mathbf{p}_{k-1}$, $\bar{\mathbf{r}}_k = \bar{\mathbf{r}}_{k-1} - \alpha_{k-1}\mathbf{A}^T\bar{\mathbf{p}}_{k-1}$, $\mathbf{x}_k = \mathbf{x}_{k-1} - \alpha_{k-1}\mathbf{A}\mathbf{p}_{k-1}$.
4 Calculate β:	$\beta_{k,k-1} = \bar{\mathbf{r}}_k^T\mathbf{r}_k/\bar{\mathbf{r}}_{k-1}^T\mathbf{r}_{k-1}$.
5 Update $\mathbf{p}$s:	$\mathbf{p}_k = \mathbf{r}_k - \beta_{k,k-1}\mathbf{p}_{k-1}$ and $\bar{\mathbf{p}}_k = \bar{\mathbf{r}}_k - \beta_{k,k-1}\bar{\mathbf{p}}_{k-1}$.
6 Convergence?	If not converged, increment k and repeat from 2.

The conjugate-gradient scheme is the same except no barred quantities are calculated and unbarred quantities replace them.

Unfortunately the same matrix symmetrization approach does not work for the minimum residual scheme, because the commutation properties are not preserved. For non-symmetric matrices one requires a **generalized minimum residual** (GMRES) method.[13] It does not symmetrize the matrix, and does not possess the currency property. It therefore needs to retain multiple older vectors to implement its conjugacy, and so requires more storage. To limit the growth of storage, it needs to be restarted after a moderate number of steps. The iterations are often called after the name of their inventor, Arnoldi. So GMRES is a "restarted Arnoldi" method. It leads to a somewhat more complicated but also quite robust scheme.

Solution of a set of *non-linear* differential equations over a large domain of N mesh points is not itself just a linear system solution. Suppose the equations for the entire domain can be written

$$\mathbf{f}(\mathbf{v}) = 0 \tag{13.32}$$

where $\mathbf{f}$ is a vector function whose N different components represent the (e.g.) finite-difference equations obeyed at all the mesh points, and $\mathbf{v}$ represents the N unknowns (typically the values at the mesh points) being solved for. The most characteristic solution method for such a system is a multidimensional Newton's method. We define the Jacobian matrix of $\mathbf{f}$ as the square $N \times N$ matrix

$$\mathbf{J}(\mathbf{v}) = \frac{\partial \mathbf{f}}{\partial \mathbf{v}}. \tag{13.33}$$

Then Newton's method is a series of iterations of the form

$$\mathbf{J}\delta\mathbf{v} = -\mathbf{f}(\mathbf{v}), \tag{13.34}$$

where $\delta\mathbf{v}$ is the change from one Newton step to the next, and $\mathbf{J}$ evaluated at the current $\mathbf{v}$ position must be used. The question is, how to solve this to find $\delta\mathbf{v}$? Well, each Newton step now *is* a linear problem, so the techniques we've been discussing apply. For the big sparse systems we get from differential equations, we want to apply an iterative scheme for each solution of $\delta\mathbf{v}$. So we are dealing with a lower level of iteration. (There are Newton iterations of iterative solutions.) If an iterative Krylov technique is used for the $\delta\mathbf{v}$ solve, then we only need to *multiply by* the Jacobian matrix; we don't actually need to form it explicitly. The multiplication of any vector $\mathbf{u}$ by the Jacobian in the vicinity of $\mathbf{v}$ can be treated approximately as

$$\mathbf{J}\mathbf{u} \approx [-\mathbf{f}(\mathbf{v} + \epsilon\mathbf{u}) - \mathbf{f}(\mathbf{v})]/\epsilon, \tag{13.35}$$

[13] A compact mathematical description is given, for example, by S. Jardin (2010), *Computational Methods for Plasma Physics*, CRC Press, Boca Raton, FL, Section 3.7.2.

where ϵ is a suitable small parameter. Therefore, rather than calculating and storing the big Jacobian matrix, all one requires, to perform a multiplication of a vector by it, is to evaluate the function $\mathbf{f}$ at two nearby $\mathbf{v}$s separated by a vector proportional to $\mathbf{u}$. This approach goes by the name "Jacobian-free Newton Krylov" method.

At the beginning of this section, the idea of iterative solution was motivated by the supposition that $\mathbf{A}$ is not very different from the identity matrix so that the natural vector update is $\mathbf{r}_k$. The speed of convergence of the iterations of any Krylov method gets faster the closer to $\mathbf{I}$ the matrix is.[14] It is therefore usually well worth the effort to transform the system so that $\mathbf{A}$ is close to identity. This process is called *preconditioning*. It consists of seeking a solution to the modified, yet equivalent, equation

$$\mathbf{C}^{-1}\mathbf{A}\mathbf{x} = \mathbf{C}^{-1}\mathbf{b}, \tag{13.36}$$

where $\mathbf{C}$ is easy to invert and is "similar to $\mathbf{A}$" in the sense that $\mathbf{C}^{-1}$ well approximates the inverse of $\mathbf{A}$. In point of fact, our general treatment already incorporated the possibility of preconditioning. The matrix $\mathbf{P}$ serves precisely the role of $\mathbf{C}^{-1}$. One does not generally multiply by $\mathbf{C}^{-1}$, and one may not even find it. Instead, the preconditioning is woven into the iterative solution by finding a preconditioned residual $\mathbf{z}$, which formally equals $\mathbf{P}\mathbf{r}$, but is in practice the solution of $\mathbf{C}\mathbf{z} = \mathbf{r}$. Therefore, the combined iterative algorithm is:

1. Update the residual $\mathbf{r}(= \mathbf{b} - \mathbf{A}\mathbf{x})$, via $\mathbf{r}_k = \mathbf{r}_{k-1} - \alpha_{k-1}\mathbf{A}\mathbf{p}_{k-1}$.
2. Solve the system $\mathbf{C}\mathbf{z}_k = \mathbf{r}_k$ to find the preconditioned residual $\mathbf{z}_k$.
3. Update the search direction using $\mathbf{z}_k$, via $\mathbf{p}_k = \mathbf{z}_k - \beta_{k,k-1}\mathbf{p}_{k-1}$.
4. Increment k and repeat from 1.

We've already encountered a preconditioner. In the bi-conjugate-gradient scheme it was formally the compound matrix $\mathbf{C}_c = \mathbf{P}_c^{-1} = \begin{pmatrix} \mathbf{0} & \mathbf{I} \\ \mathbf{I} & \mathbf{0} \end{pmatrix}$. It makes $\mathbf{C}_c^{-1}\mathbf{A}_c = \begin{pmatrix} \mathbf{A} & \mathbf{0} \\ \mathbf{0} & \mathbf{A}^T \end{pmatrix}$, which is closer to $\mathbf{I}_c$ than $\mathbf{A}_c$ is, and will be as close as

[14] The number of iterations required to converge to a given accuracy for separated eigenvalues is approximately proportional to the square root of the matrix condition number. The condition number is the ratio of the maximum to the minimum eigenvalue. That is unity for the unit matrix because all eigenvalues are 1. For a matrix representing an isotropic finite-difference Laplacian, the eigenvalues are the squared wave numbers of the spatial Fourier modes. The condition number is therefore the sum of the squares of the number of mesh points in each dimension. The number of iterations required is therefore proportional to the mesh count of the largest dimension. That is the same as optimized successive over-relation (SOR). Finding the optimum relaxation parameter for SOR is relatively easy for a rectangular mesh on a cubical domain. Consequently, un-preconditioned Krylov methods for finite-difference equations in simple domains are generally not much faster than SOR. Krylov methods come into their own when the matrices are more complicated and optimizing SOR is impossible, for example for finite-element methods on unstructured grids.

$\mathbf{A}$ is to $\mathbf{I}$. That is why the bi-conjugate-gradient scheme has the same condition number as the original asymmetric matrix $\mathbf{A}$. We might be able to do even better than this by making the compound preconditioner $\mathbf{C}_c = \begin{pmatrix} \mathbf{0} & \mathbf{C} \\ \mathbf{C}^T & \mathbf{0} \end{pmatrix}$, giving $\mathbf{C}_c^{-1}\mathbf{A}_c = \begin{pmatrix} \mathbf{C}^{-1}\mathbf{A} & \mathbf{0} \\ \mathbf{0} & (\mathbf{C}^T)^{-1}\mathbf{A}^T \end{pmatrix}$. Here $\mathbf{C}$ is the additional preconditioner chosen as an easily inverted approximation to $\mathbf{A}$.

There are many other possible preconditioners for Krylov schemes. The preconditioning step 2 might even itself consist of some modest number of steps of an iterative matrix solver of a different type, for example approximate SOR inversion of $\mathbf{A}$. In a sense, then, preconditioning interleaves two different approximate matrix solvers together, in an attempt to gain the strengths of both. In order to preserve the currency-property and avoid retaining past residuals, we must observe the commutation and symmetry requirements. Conjugate gradient schemes take $\mathbf{AR}=\mathbf{I}$ and so always commute. The only other requirements are that the preconditioning matrix (as well as $\mathbf{A}$) be symmetric so that $\mathbf{P}=\mathbf{C}^{-1}$ is symmetric (which requires that $\mathbf{C}$ be symmetric). The GMRES scheme does not possess the currency property; so it does not have to use symmetric preconditioning.

At the very least, any Krylov system should be Jacobi preconditioned. The Jacobi $\mathbf{C}$ consists of the diagonal matrix whose elements are the diagonals of $\mathbf{A}$. Jacobi preconditioning is completely equivalent to scaling the rows of the system so that all of the matrix diagonal entries become one. It is usually most efficient simply to perform that scaling rather than using an explicit Jacobi preconditioner. When all the diagonal entries are already equal, as they are for many cases where $\mathbf{A}$ represents spatial finite differences, the system is already effectively Jacobi preconditioned. Other preconditioners generally require specific sparse-matrix partial decomposition in order to minimize their costs. It is hard to predict their computational advantage. Sometimes they dramatically reduce the number of iterations, but for sparse matrices they also generally significantly increase the arithmetic per iteration.

13.4 Fluid evolution schemes

13.4.1 Incompressible fluids and pressure correction

In Chapter 7 we saw that for a fluid flow speed small compared with the sound speed $c_s = \sqrt{\gamma p_0/\rho_0}$ the CFL criterion forced us to take small timesteps $\Delta t \leq \Delta x/c_s$. So, as the sound speed gets greater (compared with other speeds of interest) an explicit solution of a certain time duration requires more and more, smaller and smaller timesteps. It becomes a *stiff system*, of the type discussed in Section 2.3, in which a large disparity exists between the scales of different modes. Now c_s increases if γ becomes large. Remember we've expressed the equation of state as $p \propto \rho^\gamma$. A large value of γ arises if the fluid in question is very *incompressible*. Formally, a completely incompressible fluid is represented by the limit $\gamma \to \infty$, in which the sound speed becomes infinite. For large γ, it takes a large increase in pressure to cause a small increase in density. Solving nearly incompressible fluid equations using an explicit approach, like the Lax–Wendroff scheme, will be computationally costly. This is a generic problem for all types of hyperbolic problems. It arises because of having to represent a large range of velocity all the way from the velocities of interest up to the speed of propagation of the fastest wave in the problem (the sound wave in our example).

However, very often we actually don't care much about the propagation of very fast sound waves. They aren't of interest, for example, in many problems concerning the flow of liquids.[15] In these cases, we don't want to have to represent the sound waves; we just want them to go away. But if we treat the fluid equations explicitly with a tolerable sized timestep, they don't go away; instead they become unstable. What do we do? We use a numerical scheme which treats the sound waves *implicitly*, rather than explicitly.

The Navier Stokes equation (7.8) is

$$\frac{\partial}{\partial t}(\rho \boldsymbol{v}) + \nabla p = \nabla.(\rho \boldsymbol{v}\boldsymbol{v}) - \mu \nabla^2 \boldsymbol{v} + \boldsymbol{F} = \boldsymbol{G}, \tag{13.37}$$

where we've gathered the advective, viscous, and body forces together into $\boldsymbol{G}$ for convenience. Linear sound waves can be derived by setting $\boldsymbol{G} = 0$ and taking the divergence of the remainder which, invoking continuity, becomes $\partial^2 \rho/\partial t^2 = \nabla^2 p$. Since an equation of state converts ρ-variation into p-variation, it becomes a simple wave equation. This observation shows that it is the left-hand side of eq. (13.37), the relation between p and $\rho \boldsymbol{v}$, that we need to make implicit in order to stabilize sound waves. If we are advancing eq. (13.37) by steps Δt in time, then an explicit scheme could be written

$$(\rho \boldsymbol{v})^{(n+1)} - (\rho \boldsymbol{v})^{(n)} = \Delta t(-\boldsymbol{\nabla} p^{(n)} + \boldsymbol{G}^{(n)}). \tag{13.38}$$

When a fluid is essentially incompressible, the equation of state can be considered[16] to be $\boldsymbol{\nabla}.(\rho \boldsymbol{v}) = 0$. If $\rho \boldsymbol{v}$ is to satisfy the continuity equation $\boldsymbol{\nabla}.(\rho \boldsymbol{v}) = 0$ at timestep $n+1$, then the divergence of eq. (13.38) gives a

[15] There are some problems in liquids where sound waves are important; just a minority.

[16] This equation arises from substituting $\rho = const.$ into the (sourceless) continuity equation, so $0 = \partial \rho/\partial t = -\boldsymbol{\nabla}.(\rho \boldsymbol{v})$. If $\nabla \rho$ is non-zero, so $\boldsymbol{\nabla}.(\rho \boldsymbol{v})$ is non-zero even though $\boldsymbol{\nabla}.\boldsymbol{v} = 0$, then the pressure equation acquires an additional term.

Poisson equation for $p^{(n)}$:

$$\nabla^2 p^{(n)} = \nabla.\boldsymbol{G}^{(n)} + \nabla.(\rho\boldsymbol{v})^{(n)}/\Delta t. \tag{13.39}$$

The pressure $p^{(n)}$ is determined by solving this Poisson equation. Note that the second term on the right would be zero if prior steps of the scheme were exact. It can be considered a correction term to prevent build up of divergence error. Once having determined $p^{(n)}$ we can then solve for $(\rho\boldsymbol{v})^{(n+1)}$. That's an explicit scheme.

To obtain an implicit scheme we want to use values for the $n + 1$ step in the right-hand side of eq. (13.38), especially for p. The problem (as always in implicit schemes) is that we don't know the new ($n + 1$ step) values explicitly until the advance is completed. They are not available for simple substitution. However, we can treat the pressure approximately implicitly by a two-step process. First, we calculate an intermediate estimate $(\rho\boldsymbol{v})^{(n*)}$ of the new flux density by a momentum equation update using the *old* pressure gradient $\boldsymbol{\nabla} p^{(n)}$ (a step that's explicit in pressure) as $(\rho\boldsymbol{v})^{(n*)} = (\rho\boldsymbol{v})^{(n)} + \Delta t(-\boldsymbol{\nabla} p^{(n)} + \boldsymbol{G}^{(n)})$. Next, we calculate a *correction* Δp to the pressure to make it mostly implicit by solving the Poisson equation $\nabla^2 \Delta p^{(n)} = \nabla.[(\rho\boldsymbol{v})^{(n*)} - (\rho\boldsymbol{v})^{(n)}]/\Delta t$. Then, we update the flux to our approximately implicit expression: $(\rho\boldsymbol{v})^{(n+1)} = (\rho\boldsymbol{v})^{(n*)} - \Delta t \boldsymbol{\nabla}(\Delta p^{(n)})$. A number of variations on this theme of "pressure correction" have been used in numerical fluid schemes.[17] When treating plasmas as fluids, for example using magnetohydrodynamics, various anisotropic fast and slow waves appear and several methods for eliminating unwanted fast waves have been developed.[18]

13.4.2 Non-linearities, shocks, and upwind and limiter differencing

When flow velocities are in the vicinity of the sound speed, a different sort of difficulty arises in treating compressible fluids numerically. It is that nonlinearity becomes important and fluids can develop abrupt changes of parameters over short length scales that are called shocks. The problem with the Lax–Wendroff and similar schemes in such a context is that when encountering such abruptly changing fluid structures they generate spurious oscillations. This happens because such second-order accurate difference schemes are *dispersive*. Waves of short wavelength experience numerical

[17] See, e.g. J. H. Ferziger and M. Perić (2002), *Computational Methods for Fluid Dynamics*, third edition, Springer, Berlin.

[18] See, e.g. S. Jardin (2010), op cit.

phase shifts arising from the discrete approximations. It is these dispersive phase shifts that cause the oscillations.

A family of methods that avoids generating spurious oscillations are those that use *upwind* differencing. These methods are one-sided finite spatial differences, always using the upwind side of the location in question. In other words, if the fluid velocity is in the positive x-direction, then the gradient of a quantity Q at mesh node i is taken to be given by $(Q_i - Q_{i-1})/\Delta x$. If, however, the velocity is in the negative direction it is taken as $(Q_{i+1} - Q_i)/\Delta x$. The upwind choice has the effect of stabilizing the time evolution, which is advantageous. They also avoid introducing spurious non-monotonic (oscillatory) behavior. But unfortunately these schemes have rather poor accuracy. They are only first order in space, and they introduce very strong artificial (numerical) dissipation of perturbations that ought physically not to be damped.

There is a non-linear way to combine higher-order accuracy with the maintenance of monotonic behavior (hence suppressing spurious oscillations). It is to *limit* the gradients so as to prevent oscillations occurring. Oscillations are measured generally in the form of the "total variation" $\sum_i |Q_i - Q_{i-1}|$. This quantity is equal to the difference between the end points for a monotonic series, but is larger if there are regions of non-monotonic behavior. A limiter method will avoid introducing extra oscillations if it never increases the total variation; i.e. it is "total variation diminishing" (TVD). When these ideas are combined with a representation of the fluid variables as being given by their average value over a cell, together with a gradient within the cell that does not necessarily imply continuity at the cell boundary, it is possible to define consistently schemes that have greater fidelity in representing abrupt fluid transitions.[19] Nevertheless, care must be exercised in interpreting fluid behavior in the vicinity of regions of variation with scale length comparable to the mesh spacing. It is always safer if possible to ensure that the fluid variation is resolved by the mesh. If it is, then the Lax–Wendroff scheme will give second-order accurate results.

13.4.3 Turbulence

Turbulence refers to unsteady behavior in fluids flowing fast enough to generate flow instabilities. Direct numerical simulation (DNS) is the most natural way to solve a problem in which all the length scales of the turbulence can be resolved. It consists of choosing a domain size large enough to encompass

[19] A comprehensive introduction to limiter methods is given in R. J Leveque (2002), *Finite Volume Methods for Hyperbolic Problems*, Cambridge University Press, Cambridge.

the largest scale of the problem, while dividing it into a fine enough mesh to resolve the smallest scale of the turbulence. The smallest scale is generally that at which viscosity μ damps out the fluid eddies. We express the order of magnitude of gradients as $1/L$: in terms of a length scale L. Then the non-linear inertial term $\nabla.(\rho \boldsymbol{v}\boldsymbol{v})$ is balanced by the viscous term approximately when $Re_L \equiv \rho v_L L/\mu$ is of order unity. Here v_L is the velocity associated with eddies of scale L, and Re_L is the "Reynolds number" at scale L. It is approximately unity at the finest fluid scale of the problem. The Reynolds number of the macroscopic flow (Re) is given approximately by substituting the typical large-scale size and flow velocity into this definition. The critical Re at which turbulence sets in is generally somewhere between a few thousand and a few hundred thousand, depending upon geometry. Typical situations can easily give rise to macroscopic Reynolds numbers of 10^6 or more.[20] Because the ratio of largest to smallest eddy-scale can therefore be very great (often estimated as $Re^{3/4}$), the DNS approach eventually becomes computationally infeasible.

Large eddy simulation (LES) seeks to moderate the computational demands through an approximation that filters out all effects smaller than a certain length scale. Fig. 13.3 helps to explain this process. The turbulence scale length can be expressed as a wave-number $k \sim 2\pi/L$, in a Fourier transform of the perturbations. The turbulence k-spectrum is artificially cut off at a value that is generally well below the physical range embodied in the turbulence, i.e. below the viscous dissipation range where $Re_L \sim 1$. This spatial filtering permits a less-refined mesh to represent the problem. The effects of the range that is artificially cut off can be reintroduced in an approximate way into the equations solved. This has most often been done in the form of an effective eddy viscosity added to the Navier Stokes equations. However, fit estimates for the magnitude of the eddy viscosity are rather uncertain.[21] In fact, the finite resolution of a discrete grid representation gives rise to an effective k cut-off in any case. If a difference scheme is used that introduces sufficient dissipation (not just aliasing and dispersion), it is sometimes presumed that no explicit additional filtering is essential.

The Reynolds averaged Navier Stokes (RANS) is an even more approximate treatment, which averages over all relevant time-scales leaving only the steady part of the flow. Therefore, essentially *all* the turbulence effects must be represented in the form of effective transport coefficients. If we ignore for

[20] A large jet aircraft may reach 10^9.

[21] See for example U. Piomelli (1999), Large-eddy simulation: achievements and challenges, *Progress in Aerospace Sciences*, **35**, 335–362.

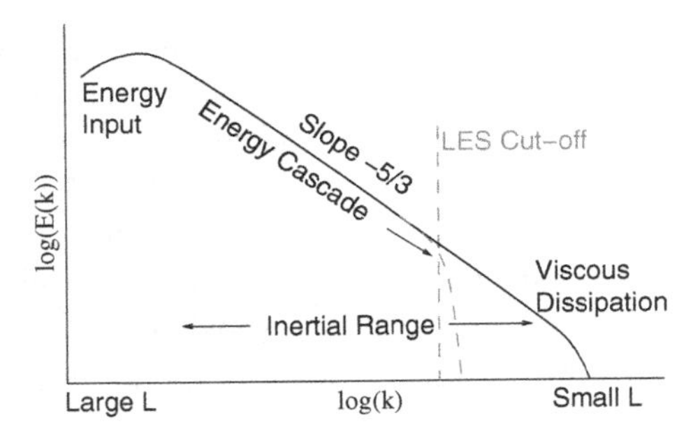

Figure 13.3 Schematic energy spectrum $E(k)$ of turbulence as a function of wave number $k = 2\pi/L$. There is an inertial range where theory (and experiments) indicate that a cascade of energy towards smaller scales gives rise to a power law $E(k) \propto k^{-5/3}$. Eventually viscosity terminates the cascade. LES artificially cuts it off at lower k.

simplicity any variation in density ρ, then when we time-average the Navier Stokes equation (13.37) all the terms except $\nabla.(\rho \boldsymbol{v}\boldsymbol{v})$ are linear. Therefore, their average is simply the average value of the corresponding quantity. The non-linear term, though, in addition to the product of the averaged quantities, will give rise to a contribution consisting of the average of the square of the fluctuating part of $\boldsymbol{v}$, i.e. the divergence of $\rho\langle\tilde{\boldsymbol{v}}\tilde{\boldsymbol{v}}\rangle$ where the over tilde denotes the fluctuating part and angular brackets denote time average. The average of the fluctuating part of a quantity is zero, but the average of the square of a fluctuation[22] is in general non-zero. This extra (tensor) term $\rho\langle\tilde{\boldsymbol{v}}\tilde{\boldsymbol{v}}\rangle$ is called the Reynolds stress. In order to solve the RANS equations, one requires a way to estimate the Reynolds stress term from the properties of the time-averaged solution. This is done by writing down higher-order moment equations and averaging them, which gives rise to an equation for the evolution of the Reynolds stress tensor. That equation contains various other terms, including a third-order tensor of the form $\langle\tilde{\boldsymbol{v}}\tilde{\boldsymbol{v}}\tilde{\boldsymbol{v}}\rangle$. The system of equations is "closed" by adopting an approximate form for the Reynolds stress evolution equation where all the terms can be evaluated from knowledge of the flow quantities and the Reynolds stress.[23] That equation is guided by theory but with coefficients fitted to experiments. These fits are called "correlations" and lead in general to a complicated tensor evolution equation. The time-averaged Navier Stokes equation and the Reynolds stress tensor evolution equation together form a composite system that is then solved numerically.

[22] Or the product of two different fluctuating quantities

[23] See for example B. E. Launder, G. J. Reece, and W. Rodi (1975), Progress in development of a Reynolds-stress turbulence closure, *Journal of Fluid Mechanics*, **68**, 537–566.

Appendix A
Summary of matrix algebra

A.1 Vector and matrix multiplication

We consider a vector $\mathbf{v}$ of length J (in an abstract vector space of dimensions J) to be an ordered sequence of J numbers.[1] The vector can be displayed either as a column

$$\mathbf{v} = \begin{pmatrix} v_1 \\ v_2 \\ \vdots \\ v_J \end{pmatrix} \tag{A.1}$$

or as a row, which we regard as the *transpose*, denoted ^{T}, of the column vector:

$$\mathbf{v}^T = (v_1, v_2, \ldots, v_J). \tag{A.2}$$

Vectors of the same dimensions can be added together so that the jth entry of $\mathbf{u} + \mathbf{v}$ is $u_j + v_j$.

The scalar product of two vectors $\mathbf{u}$, $\mathbf{v}$ in vector notation is indicated by a dot, but in matrix notation the dot is usually omitted. Instead we write it:

$$\mathbf{u}^T\mathbf{v} = \sum_{j=1}^{J} u_j v_j. \tag{A.3}$$

If we have a set of k column vectors $\mathbf{v}_k$, for $k = 1, \ldots, K$, the jth element of the kth vector can be written, V_{jk}, and they can be arrayed compactly one after

[1] Of course there are deeper ways to think about vector spaces, but we here use the simplest approach.

the other as

$$\mathbf{V} = \begin{pmatrix} V_{11} & V_{12} & \dots & V_{1K} \\ V_{21} & V_{22} & \dots & V_{2K} \\ \vdots & \vdots & \ddots & \vdots \\ V_{J1} & V_{J2} & \dots & V_{JK} \end{pmatrix}. \tag{A.4}$$

This is a matrix. We can consider matrix multiplication to be a generalization of the scalar product. So, premultiplying a $J \times K$ matrix $\mathbf{V}$ by a length J row vector $\mathbf{u}^T$ gives a new row vector of length K:

$$\mathbf{u}^T\mathbf{V} = \left(\sum_{j=1}^{J} u_j V_{j1}, \sum_{j=1}^{J} u_j V_{j2}, \dots, \sum_{j=1}^{J} u_j V_{jK} \right). \tag{A.5}$$

If we further have a set of M row vectors, we can display them as a matrix:

$$\mathbf{U} = \begin{pmatrix} U_{11} & U_{12} & \dots & U_{1J} \\ U_{21} & U_{22} & \dots & U_{2J} \\ \vdots & \vdots & \ddots & \vdots \\ U_{M1} & U_{M2} & \dots & U_{MJ} \end{pmatrix} \tag{A.6}$$

(dispensing with the transpose notation for brevity and consistency). And multiplication of the matrices $\mathbf{U}$ ($M \times J$) and $\mathbf{V}$ ($J \times K$) can be considered to give an $M \times K$ matrix:

$$\mathbf{UV} = \begin{pmatrix} \sum_{j=1}^{J} U_{1j}V_{j1} & \sum_{j=1}^{J} U_{1j}V_{j2} & \dots & \sum_{j=1}^{J} U_{1j}V_{jK} \\ \sum_{j=1}^{J} U_{2j}V_{j1} & \sum_{j=1}^{J} U_{2j}V_{j2} & \dots & \sum_{j=1}^{J} U_{2j}V_{jK} \\ \vdots & \vdots & \ddots & \vdots \\ \sum_{j=1}^{J} U_{Mj}V_{j1} & \sum_{j=1}^{J} U_{Mj}V_{j2} & \dots & \sum_{j=1}^{J} U_{Mj}V_{jK} \end{pmatrix}. \tag{A.7}$$

This is the definition of matrix multiplication. A matrix (or vector) can also be multiplied by a single number: a *scalar*, λ (say). The (jk)th element of $\lambda\mathbf{V}$ is λV_{jk}.

The transpose of a matrix $\mathbf{A} = (A_{ij})$ is simply the matrix formed from reversing the order of suffixes:[2] $\mathbf{A}^T = (A_{ij}^T) = (A_{ji})$. The transpose of a product of two matrices is therefore the reverse of the product of the transposes:

$$(\mathbf{AB})^T = \mathbf{B}^T\mathbf{A}^T. \tag{A.8}$$

[2] So a column vector can be considered a $J \times 1$ matrix and a row vector a $1 \times J$ matrix.

A.2 Determinants

The determinant of a square matrix is a single scalar that is an important measure of its character. Determinants may be defined inductively. Suppose we know the definition of determinants of matrices of size $(M-1)\times(M-1)$. Define the determinant of an $M \times M$ matrix $\mathbf{A}$ whose ijth entry is A_{ij}, as the expression

$$\det(\mathbf{A}) = |\mathbf{A}| = \sum_{j=1}^{M} A_{1j}\mathrm{Co}_{1j}(\mathbf{A}) \tag{A.9}$$

where $\mathrm{Co}_{ij}(\mathbf{A})$ is the ijth *cofactor* of the matrix $\mathbf{A}$. The ijth cofactor of an $M \times M$ matrix is $(-1)^{i+j}$ times the determinant of the $(M-1)\times(M-1)$ matrix obtained by removing the ith row and the jth column of the original matrix:

$$\mathrm{Co}_{ij}(\mathbf{A}) = (-1)^{i+j}\left|\left(\begin{array}{ccc|ccc} A_{11} & \dots & A_{1j-1} & A_{1j+1} & \dots & A_{1M} \\ \vdots & \vdots & \vdots & \vdots & \vdots & \vdots \\ A_{i-1,1} & \dots & A_{i-1,j-1} & A_{i-1,j+1} & \dots & A_{i-1,M} \\ \hline A_{i+1,1} & \dots & A_{i+1,j-1} & A_{i+1,j+1} & \dots & A_{i+1,M} \\ \vdots & \vdots & \vdots & \vdots & \vdots & \vdots \\ A_{M1} & \dots & A_{Mj-1} & A_{Mj+1} & \dots & A_{MM} \end{array}\right)\right|. \tag{A.10}$$

The inductive definition is completed by defining the determinant of a 1×1 matrix to be equal to its single element. The determinant of a 2×2 matrix is then $A_{11}A_{22} - A_{12}A_{21}$, and of a 3×3 matrix is $A_{11}(A_{22}A_{33} - A_{23}A_{32}) + A_{12}(A_{23}A_{31} - A_{21}A_{13}) + A_{13}(A_{21}A_{32} - A_{22}A_{31})$.

The determinant of an $M \times M$ matrix may equivalently be defined as the sum over all the $M!$ possible permutations P of the integers $1,\dots,M$ of the product of the entries $\prod_i A_{i,P(i)}$ times the signum of P (plus or minus 1 according to whether P is even or odd):

$$|\mathbf{A}| = \sum_{P} \mathrm{sgn}(P) A_{1,P(1)}A_{2,P(2)},\dots,A_{M,P(M)}. \tag{A.11}$$

This expression shows that there is nothing special about the first row in eq. (A.9). One could equally well have used any row, i, giving $|\mathbf{A}| = \sum_{j=1}^{M} A_{ij}\mathrm{Co}_{ij}(\mathbf{A})$; or one could have used any column, j, $|\mathbf{A}| = \sum_{i=1}^{M} A_{ij}\mathrm{Co}_{ij}(\mathbf{A})$. All the results are the same.

The determinant of the transpose of a matrix $\mathbf{A}$ is equal to its determinant: $|\mathbf{A}^T| = |\mathbf{A}|$. The determinant of a product of two matrices is the product of the determinants: $|\mathbf{AB}| = |\mathbf{A}||\mathbf{B}|$. A matrix is said to be *singular* if its determinant is zero, otherwise it is non-singular. If a matrix has two identical (or proportional, i.e. dependent) rows or two identical columns, then its determinant is zero and it is singular.[3]

A.3 Inverses

The *unit matrix* is square,

$$\mathbf{I} = (\delta_{ij}) = \begin{pmatrix} 1 & 0 & \dots & 0 \\ 0 & 1 & \dots & 0 \\ \vdots & \vdots & \ddots & \vdots \\ 0 & 0 & \dots & 1 \end{pmatrix}. \tag{A.12}$$

with ones on the diagonal and zeroes elsewhere. It may be of any size, N, and if need be then denoted $\mathbf{I}_N$. For any $M \times N$ matrix $\mathbf{A}$,

$$\mathbf{I}_M\mathbf{A} = \mathbf{A} \qquad \text{and} \qquad \mathbf{A}\mathbf{I}_N = \mathbf{A}. \tag{A.13}$$

The *inverse* of a square matrix $\mathbf{A}$, if it exists, is another matrix written $\mathbf{A}^{-1}$ such that[4]

$$\mathbf{A}^{-1}\mathbf{A} = \mathbf{A}\mathbf{A}^{-1} = \mathbf{I}. \tag{A.14}$$

A non-singular square matrix possesses an inverse. A singular matrix does not.

The inverse of a matrix may be identified by considering the identity:

$$\sum_{j=1}^{M} A_{ij}\mathrm{Co}_{kj}(\mathbf{A}) = \delta_{ik}|\mathbf{A}|. \tag{A.15}$$

For $i = k$, this equality arises as the expansion of the determinant by row i. For $i \neq k$, the sum represents the determinant, expanded by row k, of a matrix in which the row k has been replaced by a copy of row i. The modified matrix has two rows identical, so its determinant is zero, as is δ_{ij}, $i \neq j$. Now if we regard $\mathrm{Co}(\mathbf{A})$ as a matrix, consisting of all the cofactors. Then we can consider $\sum_{j=1}^{M} A_{ij}\mathrm{Co}_{kj}(\mathbf{A})$ as being the matrix product of $\mathbf{A}$ by the transpose

[3] Consider recursively expanding the determinant and subsequent cofactors, always choosing not to expand by the rows that are identical. Eventually one ends up with all 2×2 cofactors composed of the identical rows. They are all zero.

[4] A left inverse $\mathbf{A}^L$ such that $\mathbf{A}^L\mathbf{A} = \mathbf{I}$ must always be equal to a right inverse $\mathbf{A}^R$ such that $\mathbf{A}\mathbf{A}^R = \mathbf{I}$ because $\mathbf{A}^L = \mathbf{A}^L\mathbf{I} = \mathbf{A}^L(\mathbf{A}\mathbf{A}^R) = (\mathbf{A}^L\mathbf{A})\mathbf{A}^R = \mathbf{I}\mathbf{A}^R = \mathbf{A}^R$.

of the cofactor matrix, $\mathbf{A}\mathrm{Co}(\mathbf{A})^T$. So if $|\mathbf{A}|$ is non-zero we may divide (A.15) through by it and find

$$\mathbf{A}[\mathrm{Co}(\mathbf{A})^T/|\mathbf{A}|] = \mathbf{I}. \tag{A.16}$$

This equality shows that

$$\mathbf{A}^{-1} = \mathrm{Co}(\mathbf{A})^T/|\mathbf{A}|. \tag{A.17}$$

Consequently the solution of the non-singular matrix equation $\mathbf{Ax} = \mathbf{b}$ is

$$\mathbf{x} = \frac{\mathrm{Co}(\mathbf{A})^T\mathbf{b}}{|\mathbf{A}|}, \tag{A.18}$$

which for column vectors $\mathbf{x}$ and $\mathbf{b}$ is Cramer's rule.

The inverse of the product of two non-singular matrices is the reversed product of their inverses:

$$(\mathbf{AB})^{-1} = \mathbf{B}^{-1}\mathbf{A}^{-1}. \tag{A.19}$$

A.4 Eigenanalysis

A square matrix $\mathbf{A}$ maps the linear space of column vectors onto itself via $\mathbf{Ax} = \mathbf{y}$, with $\mathbf{y}$ the vector onto which $\mathbf{x}$ is mapped. An *eigenvector* is a vector which is mapped onto a multiple of itself. That is

$$\mathbf{Ax} = \lambda\mathbf{x}, \tag{A.20}$$

where λ is a scalar called the *eigenvalue*. In general, a square matrix of dimension N has N different eigenvectors. Obviously an eigenvector times any scalar is still an eigenvector, which is not considered to be different.

Since eq. (A.20), which is $(\mathbf{A}-\lambda\mathbf{I})\mathbf{x} = 0$, is a homogeneous equation for the elements of $\mathbf{x}$, in order for there to be a non-zero solution, $\mathbf{x}$, the determinant of the coefficients, must be zero:

$$|\mathbf{A} - \lambda\mathbf{I}| = 0. \tag{A.21}$$

For an $N \times N$ matrix, this determinant gives a polynomial of order N for λ, whose N roots are the N eigenvalues.

If $\mathbf{A}$ is symmetric, that is if $\mathbf{A}^T = \mathbf{A}$, then the eigenvectors corresponding to different eigenvalues are orthogonal, that is, their scalar product is zero. See this by considering two eigenvectors $\mathbf{e}_1$ and $\mathbf{e}_2$, corresponding to different eigenvalues λ_1, λ_2, and using the respective versions of eq. (A.20) and the properties of the transpose.

$$\mathbf{e}_2^T\mathbf{A}\mathbf{e}_1 = \mathbf{e}_2^T\lambda_1\mathbf{e}_1, \quad \mathbf{e}_2^T\mathbf{A}^T\mathbf{e}_1 = (\mathbf{e}_1^T\mathbf{A}\mathbf{e}_2)^T = (\mathbf{e}_1^T\lambda_2\mathbf{e}_2)^T = \mathbf{e}_2^T\lambda_2\mathbf{e}_1. \tag{A.22}$$

So, by subtraction,

$$0 = \mathbf{e}_2^T(\mathbf{A} - \mathbf{A}^T)\mathbf{e}_1 = (\lambda_1 - \lambda_2)\mathbf{e}_2^T\mathbf{e}_1. \tag{A.23}$$

If there are multiple independent eigenvectors with identical eigenvalues, they can be chosen to be orthogonal. In that standard case, the eigenvectors are all orthogonal: $\mathbf{e}_i^T\mathbf{e}_j = 0$ for $i \neq j$.

If we then take the eigenvectors also to be normalized such that $\mathbf{e}_j^T\mathbf{e}_j = 1$, we can construct a square matrix $\mathbf{U}$ whose columns are equal to these eigenvectors (as in eq. (A.4)). The matrix $\mathbf{U}$ whose columns are orthonormal is said to be an *orthonormal* matrix (sometimes just called orthogonal). The inverse of $\mathbf{U}$ is its transpose: $\mathbf{U}^{-1} = \mathbf{U}^T$. This $\mathbf{U}$ is a unitary basis transformation which diagonalizes $\mathbf{A}$. This fact follows from the observation that $\mathbf{AU} = \mathbf{DU} = \mathbf{UD}$, where $\mathbf{D}$ is the diagonal matrix constructed from the eigenvalues:

$$\mathbf{D} = \begin{pmatrix} \lambda_1 & 0 & \dots & 0 \\ 0 & \lambda_2 & \dots & 0 \\ \vdots & \vdots & \ddots & \vdots \\ 0 & 0 & \dots & \lambda_N \end{pmatrix}. \tag{A.24}$$

Therefore,

$$\mathbf{U}^T\mathbf{AU} = \mathbf{U}^T\mathbf{UD} = \mathbf{D}. \tag{A.25}$$

References

B. J. Alder and T. E. Wainwright (1957), Phase transition for a hard sphere system, *J. Chem. Phys.* **27**, 1208–1209.

R. Barrett, M. Berry, T. F. Chan, *et al.* (1994), *Templates for the Solution of Linear Systems: Building Blocks for Iterative Methods*, second edition, SIAM, Philadelphia, which is available at http://www.netlib.org/linalg/html_templates/report.html.

C. K. Birdsall and A. B. Langdon (1991), *Plasma Physics via Computer Simulation*, IOP Publishing, Bristol.

S. Brandt (2014), *Data Analysis: Statistical and Computational Methods for Scientists and Engineers*, fourth edition, Springer, New York.

S. C. Chapra and R. P. Canale (2006), *Numerical Methods for Engineers*, fifth edition, or later, McGraw-Hill, New York.

J. H. Ferziger and M. Perić (2002), *Computational Methods for Fluid Dynamics*, third edition, Springer, Berlin.

A. Hébert (2009), *Applied Reactor Physics*, Presses Internationales Polytechnique, Montréal.

R. W. Hockney and J. W. Eastwood (1988), *Computer Simulation using Particles*, Taylor and Francis, New York.

T. J. R. Hughes (1987), *The Finite Element Method*, Prentice Hall, Englewood Cliffs, NJ.

C. P. Jackson and P. C. Robinson (1985), A numerical study of various algorithms related to the preconditioned conjugate gradient method, *International Journal for Numerical Methods in Engineering* **21**, 1315–1338.

F. James (1994), *Computer Physics Communications* **79**, 111.

S. Jardin (2010). *Computational Methods for Plasma Physics*, CRC Press, Boca Raton.

B. E. Launder, G. J. Reece, and W. Rodi (1975), Progress in development of a Reynolds-stress turbulence closure, *Journal of Fluid Mechanics* **68**, 537–566.

R. J. Leveque (2002), *Finite Volume Methods for Hyperbolic Problems*, Cambridge University Press, Cambridge.

M. Luscher (1994), *Computer Physics Communications* **79**, 100.

G. Markham (1990), Conjugate gradient type methods for indefinite, asymmetric, and complex systems *IMA Journal of Numerical Analysis* **10**, 155–170.

U. Piomelli (1999), Large-eddy simulation: achievements and challenges, *Progress in Aerospace Sciences* **35**, 335–362.

W. H. Press, B. P. Flannery, S. A. Teukolsky, and W. T. Vettering (1989), *Numerical Recipes*, Cambridge University Press, Cambridge.

G. D. Smith (1985), *Numerical Solution of Partial Differential Equations*, Oxford University Press, Oxford, p. 275ff.

G. Strang and G. J. Fix (1973, 2008), *An Analysis of the Finite Element Method*, Reissued by Wellesley-Cambridge Press, Wellesley, MA.

Index

Printed in the United States
By Bookmasters